I0470861

U.S. Department
of Transportation
**National Highway
Traffic Safety
Administration**

www.nhtsa.gov

DOT HS 811 502

July 2011

State Traffic Information Systems Improvements: Promising Practices

DISCLAIMER

This publication is distributed by the U.S. Department of Transportation, National Highway Traffic Safety Administration, in the interest of information exchange. The opinions, findings, and conclusions expressed in this publication are those of the authors and not necessarily those of the Department of Transportation or the National Highway Traffic Safety Administration. The United States Government assumes no liability for its contents or use thereof. If trade names, manufacturers' names, or specific products are mentioned, it is because they are considered essential to the object of the publication and should not be construed as an endorsement. The United States Government does not endorse products or manufacturers.

1. Report No. DOT HS 811 502	2. Government Accession No.	3. Recipient's Catalog No.	
4. Title and Subtitle State Traffic Information Systems Improvements: Promising Practices		5. Report Date July 2011	
		6. Performing Organization Code	
7. Authors Robert A. Scopatz and Barbara Hilger DeLucia		8. Performing Organization Report No.	
9. Performing Organization Name and Address National Highway Traffic Safety Administration National Center for Statistics and Analysis National Drivers Register and Traffic Records Division NVS-422 1200 New Jersey Avenue .SE Washington, DC 20590		10. Work Unit No. (TRAIS)	
		11. Contract or Grant No. DTNH22-09-F-00212	
12. Sponsoring Agency Name and Address National Highway Traffic Safety Administration National Center for Statistics and Analysis National Drivers Register and Traffic Records Division		13. Type of Report and Period Covered NHTSA Technical Report	
		14. Sponsoring Agency Code National Drivers Register and Traffic Records Division, NVS-422	
15. Supplementary Notes The Contracting Officer Technical Representative for this project was Karen F. Scott.			

16. Abstract

This report highlights the major State-level accomplishments since 2005 in improving data systems used in traffic safety decision making. A nationwide assessment of traffic records system improvements solicited information from all NHTSA regions and all States plus the District of Columbia and U.S. territories. States were asked to report data quality improvement efforts taking place during the years following passage of the Safe Accountable Flexible Efficient Transportation Equity Act: A Legacy for Users (SAFETEA-LU) in 2005. The project was designed to identify improvements in crash, roadway, driver, vehicle, citation/adjudication, and injury surveillance datasets—the six major components of State traffic records systems. Improvements were sought in the data quality attributes of timeliness, accuracy, completeness, consistency, integration, and accessibility. State projects with quantitative measures showing data quality improvement are highlighted. Additional projects with qualitative evidence of data quality improvement are also described. Several States are listed as pursuing promising practices and the most effective of these are recommended for promotion to the traffic safety and traffic records community.

17. Key Word Traffic records, traffic safety, State traffic records systems, safety decision making, safety data quality		18. Distribution Statement Document is available to the public from the National Technical Information Service www.ntis.gov	
19. Security Classif. (of this report) Unclassified	20. Security Classif. (of this page) Unclassified	21. No. of Pages 100	22. Price

Executive Summary

This report highlights the major State-level accomplishments since 2005 in improving data systems used in traffic safety decision making. A nationwide assessment of traffic records system improvements solicited information from all NHTSA Regions and all States plus the District of Columbia and U.S. Territories. States and Territories were asked to report data quality improvement efforts taking place during the years following passage of the Safe Accountable Flexible Efficient Transportation Equity Act: A Legacy for Users (SAFETEA-LU) in 2005.

The project was designed to identify improvements in crash, roadway, driver, vehicle, citation/adjudication, and injury surveillance data systems—the six major components of State traffic records systems. Improvements were sought in the data quality attributes of timeliness, accuracy, completeness, consistency, integration, and accessibility. State projects with quantitative measures showing data quality improvement are highlighted. Additional projects with qualitative evidence of data quality improvement are also described. Several States are highlighted as pursuing promising practices and the most effective of these are recommended as examples that might be used by other States.

Table of Contents

Project Overview

Introduction

The purpose of this project is to assess the improvements that States have made in their traffic records systems during the years 2006 to 2010 following passage of the SAFETEA-LU legislation. Many data improvement projects during the period were funded under the Section 408 Data Grant Improvement Program, but this project reviews all data improvement projects regardless of funding sources. The assessment of improvements to State traffic records systems identified those projects that achieved significant benefits in relation to cost and might serve as examples to other States evaluating the implementation of similar projects. The goals of this study were to:

1) Identify and evaluate the traffic records improvements implemented by States from 2006 to2010.

2) Identify those projects that delivered significant improvements in a cost-effective manner; i.e., delivering savings and data quality benefits, especially when proven using quantifiable performance measures.

3) Develop a panel for the 2011 Traffic Records Forum to showcase a few States that have conducted traffic records systems projects with measurable data improvements, in hopes they may serve as examples to other States.

Method

The first phase of the project was to identify State traffic records improvement projects as potential candidates for highlighting as examples for other States. To accomplish this, a Screening Questionnaire (Appendix A) was sent to each of the 10 NHTSA regional offices for distribution among the States in their Regions particularly to identify those States that showed promising practices in their data improvement projects. In completing the questionnaires and follow-up with each State, the study team worked with State Traffic Records program managers, the Traffic Records Coordinating Committee, and project managers. Submission of completed questionnaires and project descriptions was voluntary. Each State described projects that they felt affected one or more of the six core traffic records system components listed below.

Crash Data Component
The centralized records of reportable crashes in the State along with methods of data collection, transfer, analysis, and access.

Roadway Data Component
Those databases that include State and local data such as roadway inventories, traffic volume, structures inventories, pavement management, and methods of collection, storage, analysis, and access.

Driver Data Component
The records of all driver licensing and history of convictions or other actions for the driver.

Vehicle Data Component
The records of all vehicle titles and registration.

Citation/Adjudication Data Component
The local and centralized records of traffic citations and arrests, court records of case processes, results of adjudication, and methods of collection, storage and sharing of records among law enforcement, courts, prosecutors and the interaction with the Driver Data Component.

Injury Surveillance Data Component
A series of databases covering prehospital (EMS), emergency department or outpatient, inpatient, trauma registry, hospital discharge, medical examiner, vital records, and related information.

Screening Criteria
The project team reviewed completed questionnaires as well as various State data improvement projects that were documented in TRIPRS, the system for Section 408 grant applications. The team followed up with each of the States and project managers to determine if their projects met a set of preliminary selection criteria. These screening criteria included:

1) Projects implemented or substantially advanced within the target period of 2005 to 2010.

2) Projects completed or at a reasonable level of maturity to be able to judge the level of success achieved.

3) Projects with measures of success based on indicators of data quality improvement in one or more data quality attributes of:
   a. Timeliness
   b. Accuracy
   c. Completeness
   d. Consistency
   e. Integration
   f. Accessibility

4) Where possible, projects with benefit/cost measures or other funding data.

This report documents responses to the initial screening questionnaire and responses to follow-up questions that helped to supplement the initial responses. These State responses are provided in Appendix B in each of their NHTSA Regions.

Selection Criteria
This section presents selection criteria used to evaluate the State projects and is followed by those projects that the review team determined best meet those criteria. These are the State projects that the review team considers are projects to be highlighted as promising practices and to serve as examples to other States considering the implementation of similar projects.

The project selection criteria reflect the ability of States to demonstrate project success in one or more of the following areas:

1) Documented cost savings resulting from the project;

2) Data quality performance measures of timeliness, accuracy, completeness, uniformity, integration, and accessibility;

3) Current status of the project (completed or in process);

4) Scope of the project's benefits;

5) Sustainability of the project and benefits, especially with regards to future funding; and

6) Ability to meet the objectives of the SAFETEA-LU legislation including:

> "...improve the timeliness, accuracy, completeness, uniformity, integration and accessibility of State data; to evaluate the effectiveness of efforts to make such improvements; to link these State data systems, including traffic records, with other data systems within the State; and to improve the compatibility of the State data systems with national data systems and data systems of other States to enhance the ability to observe and analyze national trends in crash occurrences, rates, outcomes, and circumstances."

As the project progressed through initial screening and follow-up, it became clear that cost savings and revenue benefits rarely were recorded and proved difficult to document for most States and most programs. However, at least two projects were documented that generated revenue that helped the State pay for the data improvement program and continue to offset the cost of managing a large data system. In general, however, criteria relating to data quality performance measures were the primary and consistently objective measures on which to base a project's success.

The review team used other facets of the projects (e.g., scope, sustainability, and meeting the objectives of SAFETEA-LU) as secondary criteria to differentiate between projects that are equally successful in generating measurable data quality improvements. For example, if each of two projects yielded a 10 percent increase in crash data quality, the project that had a larger scope or was clearly more sustainable would be considered preferable as a potential model for promotion to other States. That is not to say that the smaller or less sustainable project was not a success, but rather that the successful project with a broader scope is more likely to make a substantial improvement if adopted elsewhere.

It was an initial hope that at least one successful project would be identified for each of the six main traffic records system components and the six data quality attributes. This schema would result in a 6-by-6 matrix of successful projects; i.e., up to 36 projects barring duplication that would show an example of how each data component had been improved in each quality attribute.

In the process of screening and follow-up, it became clear that many projects affect more than one quality attribute of a system (e.g., many projects affected both timeliness and accuracy of crash data). In addition, several projects resulted in improvements in more than one system. Indeed, by default, a project that improves integration, for example, must affect two or more traffic records components.

Because of the results of the questionnaire, another criterion was added to select projects that had the broadest possible effect in terms of the number of data quality attributes improved and, where applicable, the number of traffic records system components affected. The final list of selected projects was no longer expected to be 36 even if the entire 6-by-6 matrix were filled since some projects' scopes extended to more than one system. As reported in the next section, there are gaps in the efforts to improve systems such that some portions of the matrix were underrepresented among the data improvement projects.

The final selection of highlighted projects was conducted in cooperation with a team composed of the contract staff, the NHTSA contracting officers technical representative, and representatives from the NHTSA regional offices. The reasons for selection of the highlighted project—their quantitative measures of improvement—were presented to the group and a final selection of projects was developed based on the group's comments and questions.

Highlighted Data Improvement Projects

With feedback provided by the project panel of NHTSA regional and headquarters staff, a list of data improvement projects has been highlighted with State projects that meet several selection criteria, have a large impact, and affect more than one data quality attribute. The highlighted projects are provided in Table 1 below.

Table 1. States Highlighted with Promising Practices

States	Crash	Roadway	Driver	Vehicle	Citation/ Adjudication	Injury Surveillance
Alabama	✓				✓	
Delaware					✓	
District of Columbia						✓
Florida	✓					
Illinois	✓					
Indiana	✓					
Iowa	✓	✓	✓			✓
Maryland			✓	✓		
Massachusetts						✓
Michigan	✓				✓	
Minnesota	✓					

North Carolina							✓
Oregon		✓					
Tennessee	✓						
Utah							✓
Virginia	✓						

Summary information on the highlighted States' programs is provided below in terms of which data component was improved. These represent the "best of the best" State data improvement projects that were identified in this study.

Crash Data Improvement Successes (in alphabetical order)

Alabama

The State implemented an electronic crash reporting system in 2007. The following quantifiable improvements were reported:

- Timeliness based on the average number of days from crash event to data entry is three days. Their baseline timeliness was an average of 40 days.

- Accuracy was reported in 2009 as above 98 percent; its baseline was not provided but was reported as "much lower."

- Completeness in terms of percentage of reports missing data in a key field dropped to zero in 2009, from a baseline of 20 percent.

- Consistency increased to 100 percent MMUCC compliance.

The State reports improved integration with roadway data due to improved linkage via GIS and improved data accessibility by having their crash data made available through the CARE network. Quantitative measures were not identified for these data quality attributes.

Florida

The State has initiated several crash data improvement projects, including a new crash report form, training, electronic data collection and transmission, and use of GIS. Performance measures have quantified the improvements in:

- Accuracy and Integration improvements have resulted in the State achieving a 98 percent successful mapping of crashes, supporting linkage to roadway inventory data using a statewide base map. Baseline data were not supplied, but this indicator was reported as representing substantial progress in automated linkage.

- Consistency is indicated by their report of greater than 97 percent MMUCC compliance and 100 percent SAFETYNET compliance with the new crash report form implemented on January 1, 2011.

Florida anticipates measurable improvements in timeliness and completeness based on electronic data transfer to the central crash repository from the Florida Highway Patrol (already in place) and several local agencies using TraCS (in progress). Metrics to quantify these further improvements were not available at present.

Illinois

The Crash Information System (CIS) is the statewide central repository and associated processes. Improvements were noted in the following performance measures:

- Timeliness based on an average of 15 days from event to receipt for data entry from a baseline of 25.63 average days and an average of 24.45 days to completion of location data entry.

- Timeliness overall has improved to an average of 24.45 days from data entry to availability for analysis (reflecting completion of location data entry). The baseline was greater than 90 days.

- Accuracy based on the percent critical data elements with "unknown" code is 9.81 percent. A baseline was not provided.

- Accessibility defined in terms of Web usage statistics for the Safety Data Mart for the week of April 23-30, 2010 that included 212 total visits and 10,956 total page views, etc.

Indiana

Under the general ARIES/eVCRS project, the State privatized the central management of crash reporting, established data quality standards, and strengthened feedback and cooperation with the local law enforcement community. The vendor supplies field data collection software to agencies throughout the State. The following quantitative measures of improvement were the basis for selecting Indiana's project:

- Accuracy/Completeness based on the percentage of reports failing accuracy or completeness checks reduced to 1.5 percent in 2008 from 37 percent in 2003.

- Timeliness based on an improvement in 2009 to 83 percent of reports submitted on time from a baseline of only 7 percent of reports submitted on time in 2004.

The State reported improved integration and accessibility, but these were not quantified. It should be noted that Indiana has one of two traffic records improvement projects that demonstrate a net revenue gain for the State (Alabama's citation project is the other).

Iowa

Through a series of 6 projects, Iowa has seen improvements in the following data quality performance measures:

- Timeliness has achieved a level of 95.5 percent of reports entered within 30 days in 2008, with a baseline 90 percent in 2006

- Completeness as measured by an anticipated 100 percent of reports including location coordinates by the end of 2010 (preliminary of 83% in 2010) with a baseline of 50 percent of reports in 2005.

- Accessibility measured in terms of agencies accessing crash data and analysis via CMAT reaching 333 agencies and organizations in 2009. This is a 14-percent increase over the prior year and represents a major increase from the baseline of zero in 2005.

- Linkage measured by the Iowa CODES project with 11 years of linked crash and injury surveillance data. Prior to the CODES project, these files were not linked.

Michigan

The Crash Process Redesign project has followed a phased approach to release of software and procedure upgrades. The Traffic Crash Reporting System (the software component of the redesign) has resulted in measurable improvements in the following quality attributes:

- Timeliness measured as an average of 21.57 days from event to report availability in 2009, with a baseline of average of 103.23 days in 2003.

- Accuracy and Completeness as measured by average of 1.14 errors per crash report in 2009, with a baseline of 2.63 errors per report in 2004.

The State reports increased accessibility, but quantitative measures were not available.

Minnesota

The Crash Database Interface for Law Enforcement Agencies creates links between local records management systems and the statewide crash repository at the Department of Public Safety, eliminating duplicate data entry. The State's measures of timeliness and completeness show improvement:

- Timeliness as measured by average days from crash to entry on the DPS central crash database. An improvement in 2008 of approximately 1 day over the baseline in 2007 of approximately 50 days for paper reports and approximately 20 days for electronic reports received.

- Completeness as measured by the State monitors missing (blank) data on seven key fields of the crash report. From 2007 (baseline) to 2008, the percentage of reports with a blank has decreased to 30 percent from 60 percent for these seven key fields.

Tennessee

The State has managed a series of related projects to improve crash reporting. These include monitoring of data quality by the TRCC, electronic data sharing for the Tennessee Highway Patrol, improved data entry in the centralized crash database, and online data access for law enforcement. Performance measures showing improvement are:

- Timeliness measures showed improvement in three areas, all of which showed major improvement between the baseline in 2007 and final numbers for 2009. For example, in 2007 80 percent of reports were greater than 90 days old when entered into the statewide database, while in 2009, the number of reports greater than 90 days dropped to less than 1 percent.

- Accuracy as measured by the motor carrier match rate in MCMIS that increased to 100 percent in 2009 over the baseline of 62 percent in 2007.

- Consistency improved due to a form revision with the State now collecting 95 percent of MMUCC data elements, up from 25 percent with the prior form in 2006.

- Accessibility, as measured by the percentage of local law enforcement agencies accessing data and reporting features of the central crash system, at 19 percent in 2009, up from 5 percent in 2007.

Virginia

Through implementation of the Traffic Records Electronic Data System (TREDS), the State has improved in the following data quality performance measures:

- Timeliness as measured by less than 1 day (average of 6 hours) from event to entry on the system in 2009, as compared to 184 days in 2008.

- Completeness as measured by the submission of latitude/longitude coordinates into the TREDS database, with 18,993 reports in 2009 over 12 reports in 2008.

- Integration as measured by automated entry of crash data into the driver history file at 100 percent in 2009, as opposed to zero percent in 2008.

- Consistency as measured by MMUCC compliance reaching 80 percent (up from 55%) following a 2007 report form revision.

The Virginia CODES project also shows at least 6 years (2001 to 2007) of linked crash and injury surveillance data with 2 more years currently being worked on. This is relevant to consideration of the Virginia TREDS program because the Virginia DMV has added traffic records data warehouse functionality to TREDS. TREDS now includes linked datasets in the warehouse and staff is developing additional linkages in an ongoing manner. The CODES linked data are one of the resources to be added.

Other States with Quantifiable Crash Projects
In addition to the State crash projects given above, the following States have made quantifiable improvements in the areas noted:

1) Louisiana (timeliness plus anticipated metrics in accuracy and accessibility),
2) Massachusetts (timeliness and accuracy),
3) North Dakota (accuracy and integration),
4) Pennsylvania (timeliness and accuracy),
5) US Virgin Islands (timeliness, completeness, and consistency),
6) Utah (timeliness and consistency), and
7) Wyoming (timeliness and consistency).

Roadway Data Improvement Success

Iowa
The State is implementing a redesigned geographic information management system (GIMS 2.2) that integrates crash and roadway information and makes the information more accessible to users. This integration is supported by the coordinated effort to improve collection of latitude/longitude coordinates on crash reports. The State notes that the creation of the linear reference system is complete, and the GIMS rollout was completed in 2010. Performance measures showing quantitative improvement are:

- Integration measured as the percentage of crash reports with coordinates included, increased to 83 percent (preliminary) in 2010 from a baseline of 70.6 percent in 2007.

- Accessibility as measured by use of the CMAT, which has increased 14.4 percent in 2009 versus 2008. In addition, the Iowa Traffic Safety Data Service continues to fill 120 data analysis requests per year serving approximately 45 State and local agencies.

Other States with Quantifiable Roadway Projects
In addition to the State roadway project cited above, the following States have made quantifiable improvements in the areas noted:

1) Florida (accessibility), and
2) Oregon (completeness).

Driver Data Improvement Success

Iowa
The State has improved the integration of driver and vehicle files and has increased the accessibility of driver data using linked crash, driver, and injury surveillance data. There are no formal performance measures for the effort to link the driver and vehicle databases since did not measure the degree of partial integration. It can be

said that the data components went from partial integration to complete integration because of the project. Access to linked driver, crash, and injury surveillance data has supported analyses not previously possible with relation to driver behavior change. The number of people accessing the new reports is not tracked, but the availability of such analyses has obviously increased from none prior to 2009.

Maryland

The Maryland Motor Vehicle Administration has implemented its Motorcycle Safety Program to evaluate the safety impact of rider training and ownership characteristics. This project successfully links driver, vehicle, crash, citation, and injury surveillance data for use in a multifactor analysis of differential safety outcomes for riders. The University of Maryland's National Study Center for Trauma and EMS (also the CODES contractor) is responsible for the analysis.

Vehicle Data Improvement Success

Maryland

The Maryland Motor Vehicle Administration has implemented its Motorcycle Safety Program to evaluate the safety impact of rider training and ownership characteristics. This project successfully links driver, vehicle, crash, citation, and injury surveillance data for use in a multifactor analysis of differential safety outcomes for riders. This project is the sole example of improvements related to vehicle data.

Citation/Adjudication Data Improvement Successes

Alabama

The State's electronic citation project has achieved "near 100 percent accuracy" (baseline measures were not available). Improved completeness is indicated by a projected revenue increase approaching $50 million. This is one of two projects identified that show a revenue impact (the other project was Indiana's crash data privatization).

Delaware

The State implemented a GPS location-coding tool for its electronic citation project beginning in 2008. Against a baseline of zero coordinates collected in 2007, the State shows collection of coordinates on 67,059 citations in 2008.

Michigan

The State's Judicial Data Warehouse (JDW) has improved timeliness and availability of citation and conviction data as noted in the following performance measures:

- Timeliness measured in the time from receipt of data to availability in the database was 14 hours in 2009, over the 52 hours in 2008.

- Accessibility as measured by the number of counties and courts uploading data to the JDW:

- In 2007, 60 of 83 counties up to 81 of 83 counties in 2008; and
- In 2008, 219 of 255 courts with the remaining 36 courts being certified, versus 200 of 255 courts in 2007.

Other States With Quantifiable Citation/Adjudication Projects
In addition to the State citation and adjudication projects given above, the following States have made quantifiable improvements in the areas noted:

1) Georgia (timeliness);
2) Iowa (completeness);
3) North Carolina (timeliness); and
4) Tennessee (accessibility).

Injury Surveillance Data Improvement Successes

District of Columbia
EMS run report automation began in 2006. Quantified improvements have been reported as follows:

- Completeness in terms of the percentage of EMS run reports recorded in the repository at 85 percent with 2010 performance expected to reach 100 percent, versus 21.5 percent in 2007.

- Timeliness as measured by current performance of all reports available within 10 days of the incident. There is no baseline measure.

Iowa
The State established an EMS Patient Data Warehouse (PDW) to store all prehospital patient care reports. Performance measures showing improvement are:

- Completeness in terms of percentage of runs (based on call volume) reported to the PDW at 90 percent in 2009, versus only 50 percent in 2005.

- Linkage and Accessibility in terms of Iowa CODES making 11 years of linked crash and injury surveillance data (including EMS) available to users.

- Consistency in terms of the EMS reporting system exceeding the National Emergency Medical System Information System (NEMSIS) Silver standard.

Massachusetts
The Massachusetts Trauma Registry Information System (MATRIS) captures data from all trauma cases in the State. The Massachusetts Department of Public Health is linking multiple sources of injury surveillance data (prehospital, trauma, and in-patient). The State did not supply performance measures to quantify its claims for improved completeness; however, documentation provided does show the linking variables used and examples of reports using the successfully linked data.

North Carolina

The State's centralized injury surveillance data bases are all monitored for timeliness. Although baseline measures were not provided (i.e., the State did not maintain a database of paper reports), the following measurements show the current data quality: EMS runs are reported within 24 hours of the incident; trauma registry data are provided weekly; and emergency department data is reported daily. In addition, completeness of the trauma registry information is recorded as 100 percent.

Utah

Utah implemented a centralized prehospital patient care reporting system in 2006. Performance measures showing improvement are:

- Timeliness measured by average days from event to posting on the database, at 33 days in 2009 versus 180 days in 2006.

- Consistency as indicated by the statewide EMS run reporting system described as "NEMSIS-compliant" (although the level of compliance was not reported).

- Accessibility measured by the prehospital data being accessed by 139 agencies as of 2009, up from zero prior to system implementation in 2006.

Other States With Quantifiable Injury Surveillance Projects

In addition to the State injury surveillance projects given above, the following States have made quantifiable improvements in the areas noted:

1) Arkansas (data consistency);
2) Florida (data accessibility);
3) Kansas (data consistency);
4) Missouri (data consistency and accessibility); and
5) Tennessee (data completeness).

National Review (Projects Supported by Valid Measures)

Quantitative evidence of performance improvement was found in the following systems for the data quality attributes as listed. These are marked in green in the table of national results.

Quality Attribute / System	Crash	Roadway	Driver	Vehicle	Citation/ Adjudication	Injury Surveillance
Timeliness	■				■	■
Accuracy	■				■	
Completeness	■	■			■	■
Consistency	■					■
Integration	■	■	■	■	■	■
Accessibility	■	■	■	■	■	■

Crash Data Component

The largest number of State data improvement projects involved at least the crash data component and, in many cases, more than one of the quality attributes were being measured to show project improvement. The table below is a summary of the projects to improve the crash data component and the various quality attributes that are being measured by the States. Beneath the table are the performance measures reported by the States for these projects. More complete information about each project can be found in Appendix B.

States	Timeliness	Accuracy	Completeness	Consistency	Integration	Accessibility
Alabama	✓	✓	✓	✓		
Delaware	✓					
Florida		✓		✓	✓	
Illinois	✓	✓				✓
Indiana	✓	✓	✓			
Iowa	✓	✓	✓		✓	✓
Maine	✓			✓		
Maryland			✓			
Massachusetts	✓	✓				
Michigan	✓	✓	✓			
Minnesota	✓		✓			
Missouri	✓					
Nebraska	✓					
North Carolina	✓					
North Dakota		✓			✓	
Oregon			✓			

States	Timeliness	Accuracy	Completeness	Consistency	Integration	Accessibility
Tennessee	✓	✓		✓		✓
Utah	✓			✓		
Virginia	✓		✓	✓	✓	
Wyoming				✓		

Timeliness

Alabama
Timeliness of crash data has improved from about 40 days from event to entry on the statewide database down to about 3 days due to a combination of field data collection and electronic data transfer.

Delaware
Timeliness of crash data has improved from 90 days post-crash in 2007 to 30 days in 2008.

Illinois
Data show improvement in timeliness measured in terms of number of days from crash event to data entry and average number of days from data entry to date when the data are available for analysis.

Indiana
The State provided data showing on-time submission of crash reports has gone from 7 percent in 2004 to 83 percent in 2009.

Iowa
Iowa has increased the percentage of crash reports received within 30 days from 90 percent in 2006 to 95.5 percent in 2008. Louisiana: Timeliness has improved each year from 2006 through 2009 based on measures of the percentage of crashes reported in fewer than 30, 60, or 90 days, and in the average number of days from crash even to receipt of the report for data entry. The percentage received within 30 days of the crash event was 33 percent in 2005 and has increased to 66 percent in 2009.

Maine
MCRS project includes valid measures of timeliness showing improvement.

Massachusetts
The CMV crash data improvement project has generated measureable improvement in timeliness of reporting to SAFETYNET.

Michigan
Annual average reporting days has dropped from 103 days since 2003 to 21.57 in 2009, and 15.2 days in 2010 (to date).

Minnesota
Average days from crash event to data entry for online submissions have dropped from 19 days in 2007 to 3 days (in late 2008). The State is preparing data that is

more recent. For paper submissions, timeliness has improved from 51 days to 20 days over the same period.

Missouri
MSHP crash reports are available in the STARS database within 7 days versus a baseline of 22.5 days before electronic capture and transmission were implemented. Electronic transfer of other agencies' data is being pursued through the LETS project.

Nebraska
Crash data timeliness has improved from a baseline of 120 days (from event to availability of data on the statewide system) down to 90 days or less in 2009. This is the result from a partial implementation, so more improvements are expected as the project to accept data electronically is implemented in more law enforcement agencies.

North Carolina
Timeliness of crash reporting has improved from greater than 90 days in 2004 (prior to a TraCS implementation) to fewer than 7 days in 2008.

Tennessee
Decrease in percentage of reports entered more than 90 days after the crash event: 80 percent in 2007 down to <1 percent in 2009. Increase from 39 percent in 2007 to 51 percent in 2009 for percentage of crashes entered within 30 days of the event. Also an increase from 2 percent in 2007 to 91 percent in 2009 in percentage of crashes entered by 60 days after the event.

Utah
Since 2008, the crash data backlog has been reduced to 60 days or less.

Virginia
TREDS project includes valid measures of timeliness showing improvement.

Wyoming
Wyoming has implemented field data collection and electronic data transfer and has seen the number of days from crash event to availability of data on the system drop from 83 days to 16.2 days.

Accuracy
Alabama
Accuracy has improved to an estimated error rate of less than 2 percent based on the use of electronic crash reporting software, pick lists, and error checking.

Florida
The State has achieved 98 percent success in locating crashes on the statewide base map.

Illinois
Percentage of records containing a critical error has improved.

Indiana

Accuracy of crash reports has improved from 37 percent error rate in 2003 to 1.5 percent error rate in 2008. The State is also adopting a point-and-click map interface for crash location coding that will improve the accuracy of location data on crash reports.

Iowa

Accuracy of location information has improved through use of the map-based point-and-click interface. The performance measure is percentage of electronic reports with a valid latitude and longitude coordinates based on the incident location tool (ILT).

Massachusetts

Accuracy of CMV-involved crashes as reported to SAFETYNET has improved, as documented by FMCSA.

Michigan

The error rate has dropped from 2.63 errors per crash report in 2004, down to 1.20 in 2009, and 0.75 in 2010 (to date).

North Dakota

With TraCS reporting accounting for 80 percent of all crash reports, accuracy has improved. Location data accuracy is being highlighted in conjunction with the effort described under data integration with roadway information.

Tennessee

Increase from 62 percent in 2007 to 100 percent in 2009 of motor carriers matched in MCMIS.

Completeness

Alabama

The State reports 100 percent compliance with mandatory data fields using the electronic crash reporting software. An estimate of 20 percent missing data was provided as baseline.

Indiana

Measures of completeness are improving based on edit/error checks requiring officers to review any fields coded "other" before the report is submitted as final.

Iowa

Baseline (2005) of 50 percent of electronic reports submitted with valid latitude and longitude coordinates. In 2009, 99 percent of electronic crash reports have a valid latitude/longitude. With 84 percent of crash reports received electronically in 2010 (prelim), the overall completeness of location data is also high (overall 83 percent for 2010 based on preliminary data). The expectation is that by the end of 2010, the figure will be near 100 percent.

Maryland

For CMF-involved crashes, the State has implemented a project to improve driver and vehicle information completeness. 2008 data show 79 percent driver information, up from a baseline of 72 percent, and 59 percent vehicle information, up from a baseline of 36 percent.

Michigan
Michigan monitors completeness of individual crash reports (through edit/error check processes) and compares reporting levels year-to-year at the agency level. The metrics for this latter indicator are not summarized easily, but they do serve to show whether the individual law enforcement agencies are reporting at the expected level given statewide trends.

Minnesota
Minnesota monitors the proportion of crash reports submitted by the MSP with a blank in key data fields. The State reported significant improvements in seven of the data fields between fourth quarter 2007 and fourth quarter 2008.
Oregon: The State completed a location-coding project in 2007 geocoding all crashes using latitude and longitude coordinates. One hundred percent geocoding was achieved as of 2007 data.

Virginia
TREDS project includes valid measures for completeness showing improvement.

Consistency

Alabama
The State reports 100 percent MMUCC compliance

Florida
The new crash report form (scheduled for release 1/1/2011) increases compliance to 97 percent MMUCC compliance and 100 percent SAFETYNET compliance.

Maine
MMUCC compliance increased in MCRS.

Tennessee
Increase from 25 percent in 2006 to 95 percent in 2009 of MMUCC data elements present in the crash database.

Utah
The number of data elements described as fully MMUCC-compliant has increased from 54 to 81 as of the 2008 crash report revision.

Virginia
MMUCC compliance has increased.

Wyoming
The State implemented a new crash report in 2008 that is 98 percent MMUCC-compliant.

Integration

Florida
The 98 percent success in mapping crashes means that the State can integrate crash and roadway data for all but a small minority of crashes.

Iowa

Iowa is implementing a redesigned Geographic Information Management System (GIMS 2.2). This system integrates crash and roadway data. The project is meeting its milestones, but there are no specific measures of data integration. One could look to the increase in valid latitude and longitude coordinates collected on crash reports as another indicator of success in the area of integration, since the crash data now automatically load into the map-based information system. This improvement is aided by the Linear Referencing System (LRS) that supports automated data integration from sources using any of four different location coding methods (including latitude and longitude).

North Dakota

The State converted 8 years of prior crash data from a node-based location coding system to latitude/longitude coordinate-based system. The data now integrates with its roadway data and makes it more accessible for analysis.

Virginia

TREDS project includes valid measures of data integration showing improvement.

Accessibility

Illinois

Accessibility of crash data through the Safety Data Mart has increased in terms of the number of agencies and number of users.

Iowa

The CMAT program provides access to crash data query and mapping tools to agencies and organizations. The most recent year saw a 14-percent increase in the number of users (up to 333 in 2009 versus 291 in 2008).

Tennessee

The State has seen an increase from 5 percent in 2007 to 19 percent in 2009 in the percentage of local enforcement agencies using the online system for retrieval of crash data and statistical reports.

Roadway Data Component

Four States reported data improvement projects for the roadway data component. The performance measures that were used to evaluate these projects are summarized in the table. The quality attributes of timeliness, accuracy, and consistency were not addressed specifically with performance measures. Below the table are examples of the performance measures that each State reported.

States	Timeliness	Accuracy	Completeness	Consistency	Integration	Accessibility
Florida						✓
Iowa					✓	✓
North Dakota					✓	
Oregon			✓			

Completeness

Oregon

The State has equipped 90 percent of the permanent ATR locations to collect both traffic count and speed data.

Integration

Iowa

Iowa is implementing a redesigned Geographic Information Management System (GIMS 2.2. This system integrates crash and roadway data. The project is meeting its milestones, but there are not any specific measures of data integration. One could look to the increase in valid latitude and longitude coordinates collected on crash reports as another indicator of success in the area of integration, since the crash data now automatically load into the map-based information system. This improvement is aided by the Linear Referencing System (LRS) that supports automated data integration from sources using any of four different location coding methods including latitude and longitude.

North Dakota

As Stated under crashes, the State has converted 8 years of crash data to latitude and longitude coordinates for location data in order to better integrate with roadway data.

Accessibility

Florida

The State has seen increased use of online secure systems providing user access to roadway data.

Iowa

See description above for roadway data integration. The system provides access and map-based analysis tools. In addition, the Iowa Traffic Safety Data Service managed by the Iowa State University Institute of Transpiration, serves a variety of users with integrated crash and roadway data.

Driver Data Component

Two States reported data improvement projects for the roadway data component. The performance measures that were used to evaluate these projects are summarized in the table. The data quality attributes of integration and accessibility were addressed specifically with performance measures. Below the table are examples of the performance measures that each State reported.

States	Timeliness	Accuracy	Completeness	Consistency	Integration	Accessibility
Iowa					✓	✓
Maryland					✓	✓

Integration

Iowa

The State reports that the driver and vehicle files have been integrated as part of the redesigned Driver System (DS) and Vehicle Registration and Titling (VRT) system. Completion date was 4/15/2007. Of particular note are the analyses projects using integrated driver, crash, and injury surveillance data as described under accessibility.

Maryland

The State has integrated driver, vehicle, crash, citation, and injury surveillance data as part of the Motorcycle Safety Program. The linked dataset will be used to analyze the safety impact of rider training and vehicle ownership patterns.

Accessibility

Iowa

Accessibility of driver data has improved following improvements in the driver system. Examples of access and use of the data were provided based on recent analyses conducted by the University of Iowa Injury Prevention Research Center. The projects incorporated crash, driver, and injury surveillance data components to produce.

Maryland

The Motorcycle Safety Program includes creation of a linked dataset composed of driver, vehicle, crash, citation, and injury surveillance data. Details on training experience are also included.

Vehicle Data Component

Only one State reported data improvement projects for the vehicle data component. The performance measures that were used to evaluate the project are summarized in the table. The data quality attributes of integration and accessibility were addressed specifically with performance measures. Below the table are examples of the performance measures that each State reported.

States	Timeliness	Accuracy	Completeness	Consistency	Integration	Accessibility
Maryland					✓	✓

Integration

Maryland

The State has integrated driver, vehicle, crash, citation, and injury surveillance data as part of the Motorcycle Safety Program. The linked dataset will be used to analyze the safety impact of rider training and vehicle ownership patterns.

Accessibility

Maryland

The Motorcycle Safety Program includes creation of a linked dataset composed of driver, vehicle, crash, citation, and injury surveillance data. Details on training experience are also included.

Citation/Adjudication Data Component

Two States reported data improvement projects for the citation/adjudication data component. The performance measures that were used to evaluate these projects are summarized in the table. The data quality attributes of timeliness, accuracy, completeness, and ccessibility were addressed with performance measures. Below the table are examples of the performance measures that each State reported.

States	Timeliness	Accuracy	Completeness	Consistency	Integration	Accessibility
Alabama		✓	✓			
Delaware		✓	✓			
Georgia	✓					
Iowa			✓			
Michigan	✓					✓
North Carolina	✓					
Tennessee						✓

Timeliness

Georgia

Number of days from citation issuance to transmission of a conviction to driver services has improved from the pre-2007 baseline of 180 days to 67 days in 2009 based on creation of a centralized citation data warehouse and support for electronic transmission of data.

Michigan

The Judicial Warehouse project is succeeding in making citation and adjudication data available within 14 hours of receipt (2009) versus 52 hours in 2008. The project has expanded to the majority of courts.

North Carolina

Based on the eCitation project, timeliness of data entry into the statewide citation repository (ACIS) has improved from a baseline of 8 days post-issuance in 2005 to less than 3 days in 2009.

Accuracy

Alabama

The electronic citation project has achieved "near 100 percent accuracy" according to the State. Baseline error rates were not available.

Delaware

GPS location coding on electronic citations has increased the accuracy of location data. The reporting of coordinates has increased from a baseline of zero in 2007 to 67,059 citations in 2008.

Completeness

Alabama

The State is estimating a large revenue increase based on increased citation issuance because of electronic citation issuance (eCITE). Preliminary estimates are as high as $50 million in additional revenue.

Delaware

Edit checks have improved the completeness of location data on citations. From the baseline of zero coordinates reported in 2007, the reporting of latitude and longitude coordinates has increased to 67,059 citations in 2008.

Iowa

The percentage of citations electronically submitted to the courts has increased from 25.7 percent in 2005 to 57.3 percent in 2009. This increase is due to the continued adoption of TraCS for field data collection of citation data.

Accessibility

Tennessee

The number of users subscribed to the statewide criminal justice portal has increased from 2,752 in 2007 to 5,176 in 2009. The number of traffic courts using the portal to access driver history data has increased from 2 in 2007 to 8 in 2009.

The Judicial Data Warehouse makes data available for almost all courts in a timely fashion.

Injury Surveillance Data Component

Two States reported data improvement projects for the injury surveillance data component. The performance measures that were used to evaluate these projects are summarized in the table. Below the table are examples of the performance measures that each State reported.

States	Timeliness	Accuracy	Completeness	Consistency	Integration	Accessibility
Arkansas				✓		
District of Columbia	✓		✓			
Florida						✓
Iowa				✓	✓	✓
Kansas				✓		
Maine				✓		
Massachusetts			✓		✓	
Missouri	✓			✓		✓
North Carolina	✓		✓			
Tennessee			✓		✓	
Utah	✓		✓	✓		✓
Virginia					✓	

Timeliness

District of Columbia

All EMS run reports are received and available in the system within 10 days of the event.

Missouri

The Missouri Ambulance Reporting System (MARS) tracks timeliness of submissions. A comparison to baseline (paper) reporting is not available.

North Carolina

Timeliness of EMS data is now within 24 hours of the run. Trauma data is provided weekly. Emergency department data is provided daily. Baseline measures were not provided.

Utah

With implementation of a centralized prehospital reporting system beginning 2006, timeliness has improved from 180 days to 33 days post-event.

Completeness

District of Columbia
One hundred percent of EMS runs are entered into the central database.

Iowa
Completeness of EMS run data in the EMS Patient Data Warehouse reached 90 percent of call volume in 2009, up from the baseline of 50 percent in 2005, and 81 percent in 2008.

Massachusetts
The MA Trauma Registry Information System (MATRIS) includes validation checks for data from all submitting hospitals, including strict rules on use of "unspecified" codes.

North Carolina
One hundred percent of trauma events are recorded in the central database.

Tennessee
The number of ambulance services submitting EMS run reports increased from 90 in 2006 to 189 in 2009. One hundred percent of trauma centers submit data to the EMITS statewide database.

Utah
Utah reports 139 agencies are reporting to the centralized EMS database as of 2009.

Consistency

Arkansas
The EMS reporting system meets the NEMSIS Silver standard.

Iowa
Iowa reports that its central EMS database exceeds NEMSIS Silver compliance.

Kansas
The statewide EMS database includes all 83 of the NEMSIS recommended State-level data elements. Some 192 NEMSIS-compliant data elements are captured overall.

Maine
EMS run report software is NEMSIS Gold-compliant.

Missouri
The Missouri Ambulance Reporting System (MARS) is NEMSIS-compliant.

Utah
The statewide EMS reporting system is described as NEMSIS-compliant. The State did not specify which level of compliance has been achieved.

Integration

Iowa
Iowa CODES has 11 years of linked crash and injury surveillance data.

Massachusetts

Massachusetts is linking EMS, trauma, hospital discharge, and crash data. The State provided a cross-referenced list of variables and described several ways in which the data is currently linked.

Tennessee

One hundred percent of trauma centers submit data to the statewide EMITS system.

Virginia

The VA CODES project has measures of successful integration of crash data with EMS, trauma registry, hospital discharge, and vital records.

Accessibility

Florida

The State has seen increased use of online secure systems providing user access to injury surveillance system data.

Iowa

Iowa CODES project makes linked data files available for use in various studies for the State and NHTSA.

Missouri

The MARS data is accessible in a statewide database.

Utah

The EMS Bureau's prehospital data reporting system (Polaris) is accessed by 139 agencies as of 2009.

National Review (Projects Not Yet Supported by Valid Measures)

System Quality Attribute	Crash	Roadway	Driver	Vehicle	Citation/ Adjudication	Injury Surveillance
Timeliness						
Accuracy						
Completeness						
Consistency						
Integration						
Accessibility						

The following data system improvements were self-reported by the States, but were not fully supported with quantitative measures of data quality improvement. These are marked in yellow in the above table to indicate that some support was provided for the improvement, but that, for various reasons, the support did not rise to the same level as for the improvements noted in the preceding section. The reasons for the lower designation included:

- Qualitative rather than quantitative support for the improvement, although, the qualitative information had to be considered credible and likely to be supported by an actual performance measure if one were to be created.

- The project is not yet mature enough to judge it confidently as a success. This may mean that the project is still in a pilot or preliminary stage or has some other limit on the scope of the project. If a State showed overall improvement for a system, even from a project of limited scope, the designation would have been green rather than yellow.

- The project performance measures, though quantitative in nature, do not adequately address the data quality attribute. For example, a measure that claims improvement in accuracy based on the proportion of reports generated electronically is not specific to the accuracy attribute. It is probable that an accuracy improvement was achieved, but the measurements supplied do not adequately address the issue.

- Uncertainty as to the intent of a particular project and whether or not it has had an effect on a data quality attribute. For example, if a project improved the accuracy of location data, but the State did not also claim an effect on data integration, this likely improvement is highlighted in yellow.

In the list that follows, all yellow-highlighted improvements reported by the States are presented. If a different State achieved the higher (green) level of quality improvement, then in the overall national table, the green indication would take precedent. The following list includes those "yellow" results that were superseded by "green" results in other States as well as those that represent the best results reported in the study.

State-specific results in the remainder of this section indicate improvements that were reported, but where no quantitative or qualitative evidence supports the claims of data improvement due to a project.

Crash Data Component

Quality Attribute / State	AL	LA	MI	NV	OR	UT	WY
Timeliness				✓			
Accuracy		✓		✓			

Completeness								✓
Consistency								
Integration	✓							✓
Accessibility	✓	✓	✓			✓	✓	✓

Alabama

The State implemented an electronic crash reporting system beginning in July 2007 with final rollout in June 2009. Performance measures were supplied for timeliness, accuracy, completeness, and MMUCC compliance (now 100%). Qualitative descriptions of improved internal consistency, integration, and accessibility were provided without quantification.

Louisiana

Accuracy improvements are expected based on a targeted audit process being implemented by LSU in cooperation with the law enforcement agencies. The process involves focusing on critical data elements (such as location) and selection of a random sample of crash reports from each law enforcement agency. The crash reports are examined for errors and a report is provided to the submitting agency. In follow-up, the State is in the process of creating measures of accuracy and accessibility.

Michigan

Beginning in 2002, the Crash Process Redesign has followed a phased approach to the release of upgraded software and procedures (now in Release 8). The Traffic Crash Reporting System (TCRS) and related projects have yielded major improvements in timeliness, accuracy, and completeness (as shown in the performance measures provided in the submission). There are also benefits in data accessibility and analysis, but these may not have been measured using a data quality metric.

Nevada

Nevada has doubled the number of agencies collecting crash and citation data electronically. In 2009 17 out of 36 agencies (covering >96% of the population) collect crash data and issue citations electronically.

Oregon

The adoption of Crash Magic software in 2009 has improved access to data for cities and counties. In addition, online access to crash data for local agencies has increased to roughly 10 percent of law enforcement agencies.

Utah

Utah rebuilt its electronic crash reporting system in 2006, deploying a Web-based system, the Centralized Crash Repository. From 2008 on, Utah has maintained data backlogs of 60 days or less from crash event to the date the data is available on the system. At the same time, a form revision resulted in an increase from 54 to 81 data elements in full compliance with the MMUCC guideline. The new Centralized Crash Repository also improves accessibility by supporting electronic uploads and downloads of crash data for authorized agencies, including several local law enforcement agencies. Measures of access by these agencies were not provided.

Wyoming

Wyoming implemented an electronic field data collection system for crashes. The system has improved timeliness from 83 days down to 16.2 days from date of the crash to availability of data on the system. In addition, the software includes error checks to ensure that key fields are not left blank on the form, improving completeness; this latter is not measured using a data quality metric, however.

Roadway Data Component

Quality Attribute \ State	AL	MA	NV	WY
Timeliness	✓	✓		✓
Accuracy	✓	✓		✓
Completeness		✓		
Consistency		✓		
Integration	✓	✓	✓	✓
Accessibility	✓	✓		✓

Alabama

The State implemented an electronic crash reporting system beginning in July 2007 with final rollout in June 2009. Performance measures were supplied for timeliness, accuracy, completeness, and MMUCC compliance (now 100%). Qualitative descriptions of improved internal consistency, integration, and accessibility were provided without quantification.

Massachusetts

The Massachusetts DOT has embarked on two main projects to improve the roadway inventory data and add easy-to-use mapping capabilities to the department's geospatial tools. The primary measurable improvements will be in completeness and consistency of the roadway inventory data. The mapping capability will make the data more accessible, but it may be difficult to provide a performance measure demonstrating this improvement. The roadway information is coded yellow since the project to collect inventory data on all numbered routes is a multiyear effort that has an indefinite end date. It is likely to extend far beyond the window for this evaluation. As of the follow-up with Mass DOT, they are about 10 percent complete on the 11,000-mile project.

Nevada
Nevada has doubled the number of agencies collecting crash and citation data electronically. In 2009 17 out of 36 agencies (covering >96% of the population) collect crash data and issue citations electronically

Wyoming
Wyoming implemented an electronic field data collection system for crashes. The system has improved timeliness from 83 days down to 16.2 days from date of the crash to availability of data on the system. In addition, the software includes error checks to ensure that key fields are not left blank on the form, improving completeness but this latter is not measured using a data quality metric.

Driver Data Component

State Quality Attribute	NE
Timeliness	
Accuracy	
Completeness	✓
Consistency	
Integration	
Accessibility	

Nebraska
The Douglas County Court and the Nebraska DMV established a joint process to track fail-to-pay cases and to improve collection at the local level prior to referral to DMV for suspension. In the 6-month baseline period, Douglas County referred 4,141 cases to the DMV for failure to pay fines. In the six months following implementation, referrals totaled only 213 cases, a 93-percent reduction. This project is given a "yellow" indicator because of its limited scope.

Citation/Adjudication Component

Quality Attribute / State	AL	DE	MD	MO	NE	NV
Timeliness	✓			✓	✓	✓
Accuracy			✓	✓		✓
Completeness			✓	✓		
Consistency	✓					
Integration	✓	✓	✓			
Accessibility	✓					

Alabama

The Alabama electronic citation program (eCITE program provides for field data collection and electronic submission of citation data to the courts. The eCITE software includes error checking to improve both accuracy and completeness (both stated to be "near 100%" for electronic citations). Qualitative descriptions of improved integration and accessibility based on e-Cite implementation also was provided. The State also indicated that internal consistency and timeliness are improved, but there were no measures supplied.

A series of projects managed by the Alabama DOT (ALDOT) implemented a GIS-based location and analysis system. The system includes roadway inventory and crash data so that the State has improved the ability to conduct safety analyses based on roadway features and specific spot studies. Performance measures were not supplied. Follow-up determined that while the project affects the timeliness, accuracy, integration, and accessibility of roadway data, there are no quantitative measures of the improvements.

Delaware

The Delaware Criminal Justice Information System (CJIS, a.k.a. DelJIS) has completed deployment of an electronic citation module for law enforcement and the courts. CJIS also sends data electronically from the courts to the Division of Motor Vehicles. Completeness is measured in terms of the number of citations with GPS coordinates included in the electronic record. This has increased from zero in 2007 to 67,059 citations in 2008. It is also possible that the State intended to use the presence of GPS coordinate data in the citation record as an indication of its ability to link citation and crash data.

Maryland

The State is using an electronic citation system (E-TIX) that is capable of capturing GPS coordinates as part of the location information. Completeness of GPS reporting went from a baseline of 0.71 percent in the first half of 2008 to 13.6 percent through early 2009. Follow-up with the State is required to obtain data that is more recent and to clarify the calculation of the completeness metric.

Missouri
The Judicial Case Management System – Justice Information System (JIS) has been implemented in all 115 State courts. This program establishes electronic transmission of conviction data from all courts and reduces the lag time between conviction and posting to the driver record. As most traffic cases are handled at the municipal court level, the project's impact will be most obvious as the JIS is rolled out to the municipal courts – a multiyear process that is still in the early stages.

Nebraska
The Nebraska Crime Commission is managing a project to expand the ongoing citation automation program and the Statewide Citation File. The performance measure shows that prosecutors receive traffic citation information within two days of issuance compared to a baseline of 14 days; however, scope of the project appears to be limited to a local jurisdiction.

Nevada
Nevada has doubled the number of agencies collecting crash and citation data electronically. In 2009 17 out of 36 agencies (covering >96% of the population) collect crash data and issue citations electronically.

Injury Surveillance Data Component

Quality Attribute \ State	GA	MO	OR	VA
Timeliness				✓
Accuracy		✓		✓
Completeness			✓	✓
Consistency			✓	✓
Integration		✓	✓	✓
Accessibility	✓			✓

Georgia
The Online Analytical and Statistical Information System (OASIS) has been expanded to allow public access to emergency department data for persons injured in motor vehicle crashes. Accessibility has been measured by the number of downloads of data from the site. Follow-up is required to obtain a baseline and to see if information is available on the types of people using the site (i.e., is there

increased volume and variety of uses, or has the site seen increased use mainly from previous customers?).

Missouri

The Missouri Ambulance Reporting System (MARS) is a NEMSIS-compliant EMS run report system composed of a centralized database and field data collection system for use by EMS providers. The State has measured timeliness and compliance with the NEMSIS standard, but the system should improve accuracy, integration, and accessibility as well.

Oregon

The completeness of the statewide EMS/Injury Surveillance database has improved, as evidenced by the increase, from zero records in May 2008 to 24,089 records in January 2009, in the number of EMS run reports populating a pilot test database of run reports from May 2008 supplied by 19 service provider agencies.

The uniformity of the statewide EMS/Injury Surveillance database has improved, as evidenced by the increase, from zero records in May 2008 to 24,089 records in January 2009, each with 60 NEMSIS-compliant elements, in the number of EMS run reports populating a pilot test database of run reports from May 2008 supplied by 19 service provider agencies.

A project was started in 2008 using ImageTrend software to demonstrate the feasibility, challenges, and benefits of linking EMS patient records to hospital outcomes and other existing databases in the State (e.g., Oregon State Trauma System Database, Oregon State Hospital Discharge Database, ODOT Statewide Crash Database, Medical Examiner data). Data analysts were able to identify and describe which prehospital variables were most critical for matching and linking records with the EMS patient encounter database. Project deliverables included a summary document detailing a vision and general mechanics for development of a State EMS database linked to hospital outcomes.

Virginia

The Virginia Department of Health's Office of Emergency Medical Services has developed a Web-based reporting system that is NEMSIS-compliant for State-level data reporting. Individual EMS providers are given access to their own data through the Web portal. The system is described as improving timeliness, completeness, accuracy, and consistency, but no metrics or performance measures were supplied or described. Based on follow-up with the State, all of the performance improvements described for the EMS run reporting system are qualitative, not quantitative.

Discussion and Conclusions

Several States reported quantifiable improvements in just one data quality attribute (typically timeliness). Many claims of crash system improvements are supported by qualitative (nonquantitative) information. This information, though anecdotal in nature, may also serve to show that States have concentrated extensively on crash data improvement, regardless of whether or not they were able to quantify the improvements they made. In the words of one State project manager:

"We had to choose between measuring and actually making the improvements. We chose to make the improvements."

This sentiment was common among the State project managers. At least at the beginning of the 5-year period under consideration in this report, many of the managers were new to the concept of data quality measurement. A careful review of the performance measures produced by the States shows more evidence of the overall novelty of data quality measurement for most States. When asked to produce measures, the results for many of the States resulted in:

1) An inability to produce any significant measures of data quality;
2) Measures that are based on only convenience and/or Federal requirements (e.g., accuracy measured based on SAFETYNET reporting standards);
3) Measures that are focused narrowly (e.g., completeness measured by the percentage of reports with data in one or two data elements such as latitude/ longitude coordinates).

This report highlights, from among many, those projects that have the most compelling evidence of improvement in a component of the statewide traffic records systems. It is clear from the information provided by the States that improvements have been achieved and that the focus areas have been primarily in crash systems and, to a lesser extent, injury surveillance systems. The latter system projects are dominated by projects establishing EMS run reporting and trauma registry databases for the first time.

With respect to the selection criteria for a promising practice, the selection can be made most accurately for crash reporting. The system improvements in crash data are well documented and, in many cases, clearly quantified. This may be an artifact of having so many projects to choose from, but it is worth noting that States with excellent documentation of crash system improvements are often less able to document their improvements in other areas. This may indicate that States have put a greater emphasis on improving the crash data systems than other components of the traffic records system.

This may be true for a variety of reasons, including institutional barriers, primary system functions, and the cost to update some of the older, legacy systems. Additionally, the crash data are the key source of information for all safety-related analyses by stakeholders in a State. Thus, it is to be expected that this data component will receive a great deal of attention from a broad group of collectors, managers, and users. Conversely, the focus on other key data components may be less directly related to safety and/or may have a more narrow focus among collectors, managers, and users. For example, roadway, driver, and vehicle databases have a primary use that is other than safety. The safety relevance of injury surveillance systems is well recognized, but these systems are still considered from a more narrow perspective in most States than are the crash data.

Many of the performance measures provided by States for this project review fall short of the ideal expressed by NHTSA in its guidance on this subject (including the Section 408 grant submission guidelines, the training provided on www.trafficrecords101.net and other resources). The main deficiencies noted for this project are:

Lack of baseline measures

In many cases, a project was implemented and there is no sensible baseline other than the "zero" starting point. In other cases, projects were measured only in terms of their current performance, regardless of whether or not a baseline could be measured. It is thus difficult to quantify improvement, although it is abundantly clear that improvement has been achieved.

Indirect measurement

Many of the measures of performance are only tangentially related to the data quality attribute under consideration. For example, many States have combined accuracy and completeness measures—they are unable to separate errors of omission and commission. In other cases, many States have a combined measure for timeliness and accessibility such that the State has no separate measure of use of the data beyond *when* it becomes available to users.

Static measurements

Some of the performance measures are structured to be static over long periods. This is especially true of those that measure compliance with standard data definitions such as MMUCC, SAFETYNET, or NEMSIS. As such, metrics that measure consistency will show a sudden jump between the baseline level and the new level and then remain stable until the next major change in a data collection form or database. Generally, year-to-year progress is not measured in these cases.

Because of the emphasis on mature (completed or nearly completed projects) in this study, many potentially worthwhile projects could not be highlighted as proven successes. It is likely that many of these projects will qualify as "promising practices" in the future, given sufficient time and a continuation of current levels of effort. For a major project to have reached full implementation by 2009 would, in most cases, mean that it was started no later than 2007, funded no later than 2006, and initially conceived some time before that. This emphasis on projects conceived early in the period since passage of SAFETEA-LU has implications for performance measurement as well. For the most part, system-level performance measures as they currently are understood were not widely used in 2005. Many projects were started and then only later were they measured with the goal of demonstrating their impact on data quality. This means that there was a general lack of baseline measurement (as previously noted). It also means that the measures used to describe the success of a particular project may be the first-ever attempted by a State to monitor that particular aspect of data quality.

Lacking experience, State development of performance measures often failed to provide the kind of data that could support a claim for success or to quantify accurately an improvement in a system. For example, measures that track the number of agencies submitting data electronically to a system may be adequate for managing the rollout of a new automated data sharing process, but they do a poor job of measuring actual system improvement such as timeliness or accuracy. Often such measures fail to compare current participation to the overall participation possible (e.g., if all agencies were in compliance) and they often fail to track accurately to true work volume. Contributions from 100 small agencies may amount to either a large or a small proportion of overall crashes, citations, etc. With lack of knowledge or direction, little was done on projects to develop quantitative performance measures in the initial years subsequent to the passage of the SAFETEA-LU legislation. As noted, these are precisely the projects that would have reached maturity in time for inclusion in this report.

The good news is that there are a number of successful projects and that there are several States with well-developed performance measurement systems. The most noteworthy of these are in Michigan, Iowa, Alabama, and Tennessee. These four States effectively maintain ongoing performance measurement systems with a systemwide focus. These States did not need to create new performance measures for the projects they implemented, but rather were able to integrate their new project monitoring tasks into an existing framework of performance measurement.

Given these examples, it is instructive to look at where these leading States failed to develop quantitative measures of performance improvement. Working with limited resources, States concentrated on making a system improvement rather than investing in measuring their system inadequacies. This approach, though lamentable from the point of view of providing proof of their accomplishments, is sensible from the point of view of putting a State's resources where the biggest impact can be made. While reviewing State systems improvements, it was often obviously reasonable to agree with the State that a project represented an improvement even when it could not be demonstrated based on a quantifiable measure of performance.

Examples of improvements that were not quantifiable for this study include several projects where a State went from **not** having a capability to having it in essentially complete form (e.g., linking driver and vehicle data, creating the first-ever EMS run reporting system, etc.). However, other examples go beyond that simplistic type of one-time leap forward. In many cases, the States simply felt they did not have a reasonable way to quantify the improvement they hoped to achieve. As an example, the number of years of linked data in a CODES project generally increases by one as a new year's data becomes available to the analysts. That is easily quantified, but ultimately not meaningful in terms of data integration or accessibility. Better measures would be the proportion of records that can be matched between any two data sources and the number of data sources that are part of the linked dataset. Even more importantly, however, is the question of whether or not the State's decision-making benefits from the linked data and if so, in what ways. The mere presence of linked data is not as important as how those data are used to improve safety. No State has developed a good way to measure the effect of linked data on decision-making, although the motorcycle safety project in Maryland shows initial promise in this regard.

In general; however, the States have yet to come up with ways to measure the importance of their data integration or accessibility improvements with respect to safety. Thus, this report comes at a point in time when, for much of traffic records systems, data quality measurement and management is in its infancy. States still are discovering the best ways to measure improvements. The measures presented in this document reflect this early stage of development. They serve as a good baseline against which to measure progress in traffic records performance measurement in years to come. It is certain that if this snapshot had been taken in 2005 rather than 2010 the number of performance measures would have been much smaller – perhaps only available from one or two States. The progress in the past five years in the area of using performance measurement has been enormous. Five years ago, most States would have been unable to find measures beyond a global measure of crash data timeliness. Looking forward, it is likely that traffic records data improvement projects will include better measurement of its impact on data quality.

In conclusion, the States have made *significant* and *measureable* progress in several components of their traffic records systems. They have also achieved credible (and creditable) progress even when they sometimes lack strong quantifiable proof of the data quality improvements. The breadth of data quality improvement in terms of the number of traffic records system components affected was impressive. The review team anticipated that many States would show improved crash records systems, especially in terms of timeliness and consistency, because of the ongoing national focus on these two quality attributes, including NHTSA's MMUCC effort to provide a guideline for the content of crash records systems. The fact that examples of quantifiable improvement in all six of the quality attributes can be identified for crash records is an indication that States are working to improve this critical data system. The level of improvement in Citation/Adjudication and Injury Surveillance Systems is also important and indicates that States are recognizing the value of these data sources for use in identifying potential problem drivers earlier (through citation activity, not only convictions) and the value of medical outcome data in describing the consequences and risks associated with crashes. NHTSA has focused on these systems through the efforts to promote creation of citation tracking and the Model Impaired Driving Records Information System (MIDRIS) and the NEMSIS.

States recognize the value of the other traffic records data systems as well—roadway, driver, and vehicle—but that States may reasonably face barriers to improving these systems, especially the driver and vehicle data systems. These are the most often described as "legacy" systems—those based on older technology that is both costly to maintain and difficult to access. States typically tend to keep these systems in operation for decades without replacement or major modification. Many States are modernizing their roadway data systems, especially in response to new national initiatives such as the Highway Safety Manual and the Model Inventory of Roadway Elements (MIRE). Driver and vehicle records systems are large, have a primary purpose other than safety, and cost multiple millions of dollars to replace. It is therefore not surprising that these systems may lag somewhat in terms of improvement and in the States' abilities to measure improvements.

Overall, it should be recognized that SAFETEA-LU has indeed promoted data quality improvement and that it has been used effectively as a driving force in quantifying those improvements. Based on the information received from the States in this study, it is also clear that the States continue to add to their abilities to quantify data quality and thus should be increasingly capable to demonstrate improvements in the future.

Management "by the numbers" has become a common practice in traffic records—much more the rule rather than the exception. In 2005, when SAFETEA-LU became law, the opposite was the case.

Appendix A: Screening Questionnaire

Screening Questions for the States on Traffic Records System Improvements
The following items relate to ANY projects aimed at improving traffic records components during the most recent 5-year period (from October 2005 to present). We are interested in projects *regardless of how they were funded* (408 funds, other Federal, State funds, or a combination of sources). We are also interested in any size project – local, statewide, regional. The primary focus is on completed and mature (soon to be completed) system improvement efforts. If you have ongoing projects you wish to highlight, you may do so, but we ask that you limit yourself to projects that have already demonstrated a benefit as indicated by your data quality performance measures.

System Component x Data Quality Measurement Matrix
Below is a 6x6 table showing the main traffic records system components (columns) and the six main data quality attributes (rows) as discussed in SAFETEA LU and the Section 408 grant program. Please place a check mark in each cell of the table where your State or any local agency in your State has seen improvement during the past five years. NOTE: Improvement can be defined numerically (i.e., actual measurements of system performance) or qualitatively (i.e., collectors, managers, and users of the system agree that things have improved).

Quality Attribute \ System	Crash	Roadway	Driver	Vehicle	Citation/ Adjudication	Injury Surveillance
Timeliness						
Accuracy						
Completeness						
Consistency						
Integration						
Accessibility						

* The data quality attributes may be defined in your own terms or using your own State's data quality performance measures.

Please provide a brief explanation of each checkmark that you placed in the table. Improvements in this list do not necessarily have to relate directly to a system improvement project – we are interested in all improvements that you (and the collectors, managers, and users of the data) have noted in the various components. If you provide only qualitative information, however, please explain specifically why you think the system's performance has improved.

For Question 2, if you have documented the relevant projects on the TRIPRS Web site, you can respond by pointing us to the projects you wish to highlight. **_Please make sure the project contact information is correct before you use this option_**. If you have not used TRIPRS, or for any additional projects not already described in TRIPRS, please use the format provided in the question.

SEND RESPONSES BY NOVEMBER 30 TO: bscopatz@data-nexus.com

System Component Improvements
Please list any successful traffic records system improvement projects during most recent 5 years (October 2005 through present). NOTE: We are interested in projects that improved any of the components of a traffic records system:

  a. Crash data,
  b. Roadway data (inventory, traffic counts, etc.),
  c. Driver data (licensing, driver history, driver control processes data, etc.),
  d. Vehicle data (registration and titling, commercial vehicle registration, etc.),
  e. Citation and djudication (citation issuance, citation tracking, court records, etc.), and
  f. Injury surveillance (EMS, trauma registry, ED- or hospital-discharge data, etc.).

Please use the following format for each project that you list:

Project Title	
Lead Agency(-ies)	
Contact Person	
Project Description	
Current Status	
System(s) Affected	
Performance Measures	

NOTE: Please be sure to include any projects that improved consistency (e.g., MMUCC or NEMSIS compliance), data integration (electronic data transfer *or* creation of new merged datasets), and any projects that improved data availability (e.g., users access to data extracts *or* improved access to analytic resources)

Regional/Elsewhere
Please provide a note about any systems improvements that you have heard of in your region or elsewhere in the United States that sound particularly successful. Please provide a contact or agency name if you have that information.

SEND RESPONSES BY NOVEMBER 30 TO: bscopatz@data-nexus.com
Or <xxx your e-mail here

Appendix B: State Review by Regions

Appendix B documents the responses from all States that include specific projects, notes, and other information provided to the project team by each State and/or region. The States and their projects are listed within each of their respective NHTSA regional designations.

A summary table is provided for each State showing the traffic record system component by the data improvement attribute. The color codes for these table cells are as follows:

> **Green**
> The project team agreed with the State that they could produce quantitative evidence of a data improvement.
>
> **Yellow**
> The project team judged the evidence of the data improvement to be credible, even though the measure was qualitative in nature.
>
> Uncolored
> A project was reported but credible evidence of the data improvement was not supplied by the State.
>
> Unchecked Cells
> No projects were submitted for consideration.

Notes follow each table to describe the subscripts assigned to each project in the table. If the State provided examples of their metrics, data quality reports, or other information, it was included in the notes.

Appendix B: State Review by Regions

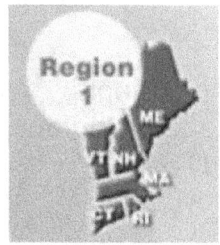

Region 1: Connecticut, Massachusetts, Maine, New Hampshire, Rhode Island, Vermont

Maine and Massachusetts responded.

MAINE

Maine's response includes the following table for measurable performance improvements – see the notes below the table for explanations of the research team's observations and opinions:

Quality Attribute \ System	Crash	Roadway	Driver	Vehicle	Citation/ Adjudication	Injury Surveillance
Timeliness	X^1					X^2
Accuracy						X^2
Completeness						X^2
Consistency	X^1					X^3
Integration						
Accessibility						

Notes:

1) The revision to MCRS includes an increase in MMUCC compatibility. As a result, the check mark for "consistency" under the crash system was added in the review. The timeliness improvements likely will start with new deployments in 2010.

2) Electronic EMS run reporting should yield measureable accuracy improvements, so the review resulted in the addition of a check mark under accuracy. Contact with the State shows that the State is measuring accuracy and likely will see improvement in the 2010-2011 timeframe. Timeliness and completeness are measurably improved and the project is mature (100% compliance in reporting was achieved in 2009).

3) Electronic EMS is listed as compliant with NEMSIS, but the level of compliance was not listed in the initial submission (e.g., Gold standard, Silver, other). As of June 2010, the EMS run report software is by ImageTrend and it is NEMSIS Gold-compliant.

Appendix B: State Review by Regions

MASSACHUSETTS

The Massachusetts response was submitted in several separate documents. The following table compiles this information into a single presentation.

System Quality Attribute	Crash	Roadway	Driver	Vehicle	Citation/ Adjudication	Injury Surveillance
Timeliness	X[1,2]	X[6]			X[4]	
Accuracy	X[1,2]	X[6]				
Completeness	X[2]	X[6]				X[5]
Consistency		X[6]				
Integration		X[6]				X[5]
Accessibility	X[3]	X[6]			X[3]	

Notes:

1) A series of projects related to Commercial Motor Vehicle (CMV) were implemented using FMCSA funding (no Section 408 grants). The projects have focused on timeliness, accuracy, and analysis of CMV-involved crash data. The projects are either completed or in process of implementing recommendations arising from those projects. Performance measures are based on the FMCSA data quality metrics for timeliness and accuracy. Based on those measures, the State has improved. The CMV project for crash data improvement has generated measureable improvements in timeliness, accuracy, completeness, and "overall" quality as measured by the FMCSA. These are well documented, but affect only a portion of the overall crash data in the Commonwealth. No follow-up documentation was provided regarding the improvements resulting from the more global RMV project.

2) The Registry of Motor Vehicles engaged in a series of projects for crash data improvement using electronic submission, Web-based submission, and follow-up with individual agencies at the municipal level. These projects have produced measureable improvements for the subset of agencies involved (i.e., those submitting data electronically and those targeted for non-submission). The RMV did not supply statewide summary data on performance measures so it was not possible to assess the overall impact of these projects. In addition, the Web-based reporting system for crashes was slated for full rollout by the end of December 2009, so no status report was provided.

3) The Executive Office of Public Safety and Security (EOPSS) funded development of a Web-based portal -- the Massachusetts Traffic Records Analysis Center (MassTRAC) for access to traffic safety data and analytic tools. Capabilities include query, filter, mapping, cross-tabulations, and reporting. The first phase provides access to the Highway Safety Division staff in EOPSS, with planned rollout to additional users in the future. No status report was provided on MassTRAC to determine how broadly it is serving the users.

4) The Merit Rating Board, the Registry of Motor Vehicles, and the Administrative Office of the District Courts have developed an electronic data transfer process for adjudicated criminal traffic violations. As of December 2009, adjudicated criminal violations are posted to the driver record automatically based on submissions from the district courts. Timeliness has improved from 33 days to 11.5 days for this subset of violations. Citation and adjudication are indicated in "yellow" in the value for timeliness based on follow-up with the MRB regarding the project to improve reporting of judgments in criminal traffic cases. The project has produced measureable improvement, but it is in fact limited to a subset of all traffic cases and does not affect all courts that hear criminal traffic cases.

5) The Massachusetts Department of Public Health is working to integrate the EMS run report and trauma registry data. The performance measure supplied by the State does not appear to relate directly to the project description supplied – data show number of trauma registry records, not a measurement of linkage. Preliminarily, the State earned a check mark under data integration for the injury surveillance data. The Department of Public Health provided information on the MATRIS system that supports completeness and integration indicators for the Injury surveillance system. The trauma registry requires validation checks from all submitting hospitals and has strict rules on use of "unspecified codes." The Integration indicator is based on linkage between EMS, Trauma, Hospital Discharge, and MMUCC variables. A cross-referenced list of variables was provided that indicated several ways in which the data are being linked now.

6) Massachusetts DOT has embarked on two main projects to improve the roadway inventory data and add easy-to-use mapping capabilities to the department's geospatial tools. The primary measurable improvements will be in completeness and consistency of the roadway inventory data. The mapping capability will make the data more accessible, but it may be difficult to provide a performance measure demonstrating this improvement. The roadway information is coded yellow since the project to collect inventory data on all numbered routes is a multiyear effort that has an indefinite end date. It is likely to extend far beyond the window for this evaluation. As of the follow-up with Mass DOT, they are about 10 percent complete on the 11,000-mile project.

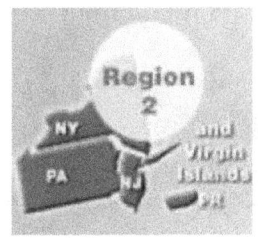

Region 2: **New Jersey, New York, Pennsylvania, Puerto Rico, U.S. Virgin Islands**

New Jersey, Pennsylvania, and the U.S. Virgin Islands responded

NEW JERSEY

System Quality Attribute	Crash	Roadway	Driver	Vehicle	Citation/ Adjudication	Injury Surveillance
Timeliness						X[1]
Accuracy						X[1]
Completeness						X[1]
Consistency						X[1]
Integration	X[1]					X[1]
Accessibility						X[1]

Notes:

1) The electronic patient care report (ePCR) project (EMS Charts) is designed to collect run reports from all EMS providers for all runs in the State. The only performance measures supplied (or in TRIPRS) relate to the number of patient care reports entered into the statewide data warehouse.

PENNSYLVANIA

System Quality Attribute	Crash	Roadway	Driver	Vehicle	Citation/ Adjudication	Injury Surveillance
Timeliness	X[1]					
Accuracy	X[1]					
Completeness	X[1]					
Consistency	X[1]					
Integration	X[1]					
Accessibility	X[1]					

Notes:

1) Pennsylvania's Crash Data Law Enforcement Liaison Program (CDLELP) is designed to improve coordination between the data managers and data collectors while promoting improved quality, adoption of electronic field data collection, and best practices. The project is believed to have improved every aspect of data quality. Based on the initial submission, the State has implemented performance measures for timeliness, accuracy, and availability (although the latter is primarily a timeliness measure as shown in the following:

Average Process Days
(Measured as average days between crash date and crash data availability to end users)

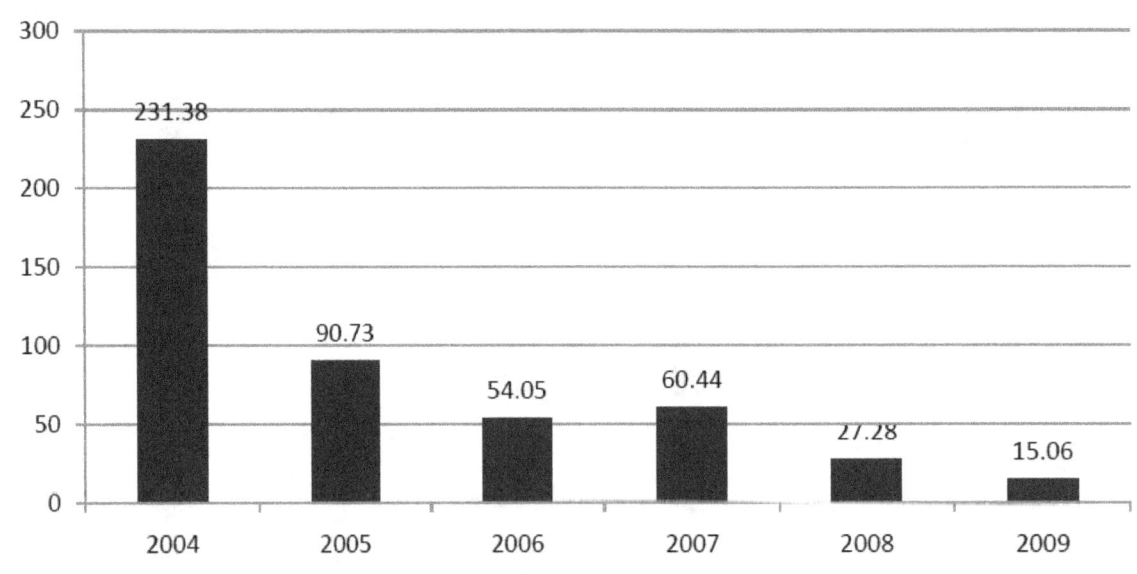

Average errors per case have also improved dramatically:

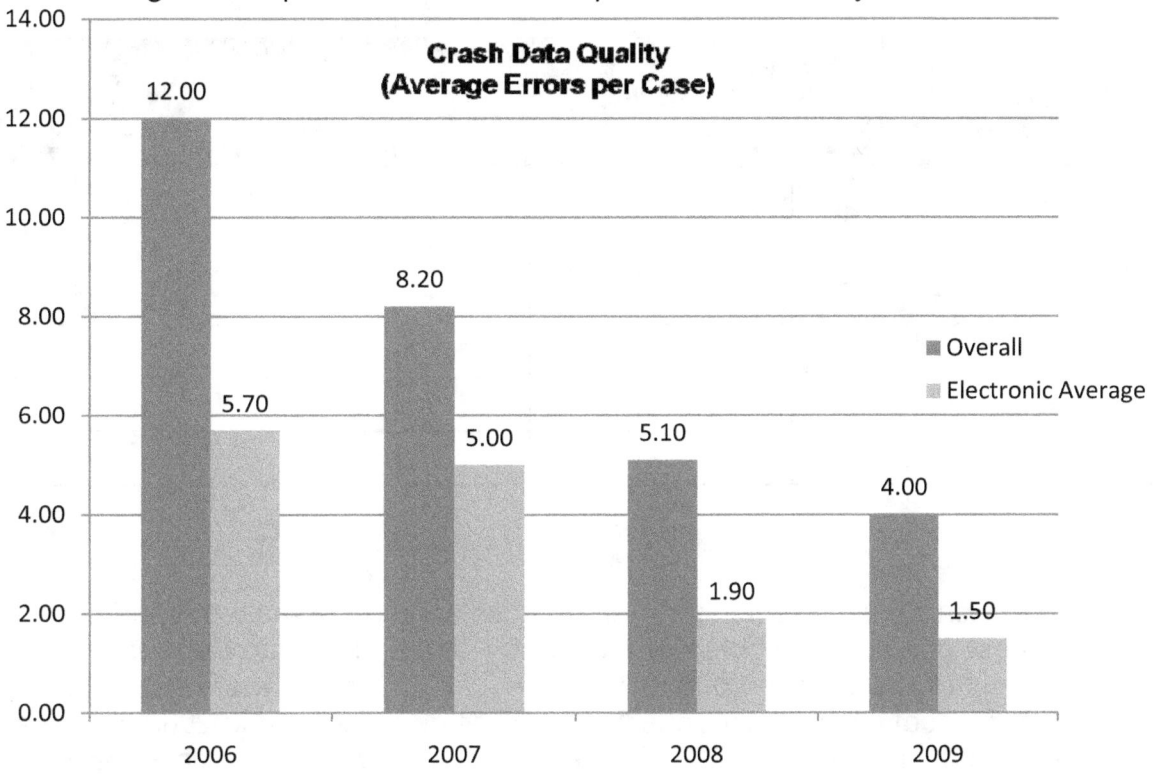

In addition, access to data for end users has become more timely as shown by indicators of access for government end users (now at 15 days post-crash down from "over 30 days" in the baseline period), and release of final fatality numbers (down from 120 days after the year end to 43 days after the yearend).

Appendix B: State Review by Regions

US VIRGIN ISLANDS

System Quality Attribute	Crash	Roadway	Driver	Vehicle	Citation/ Adjudication	Injury Surveillance
Timeliness	X^1					X^3
Accuracy	X^1					X^3
Completeness	X^1					X^3
Consistency	X^2					X^3
Integration	X^1					X^3
Accessibility	X^1					X^3

Notes:

1) The electronic crash reporting system has replaced paper-based reporting. Timeliness of data (from crash event to entry on the statewide database) has improved from 30 days in 2006 to 7 days in 2010. Completeness has improved to 93 percent of electronic reports having all fields completed in 2008 versus a baseline of 31 percent in 2007 – 77 percent of all reports are complete (including paper and electronic reporting). Other measures of performance were not quantified, but by self-reporting, the accuracy, integration, and accessibility have improved as well.

2) The USVI traffic crash report was revised in 2007 to include additional MMUCC-compliant data elements and attributes. The form went from 11 full MMUCC elements, 34 partial elements, and 94 attributes to 14 full elements, 44 partial elements, and 200 attributes.

3) The Virgin Islands Patient Care Reporting System (VIPCR) is an electronic EMS run report form reported as NEMSIS-compliant. The system includes a centralized database capable of storing data submitted territory-wide. Reporting to this system has increased from no traffic cases prior to 2009 up to 60 percent of all traffic-related runs reported in 2009. The territory indicates that all data quality attributes were improved by this system.

4) As of June 2010, the EMS reporting system is in a gradual rollout (60% reporting to date) so is neither mature or complete.

Appendix B: State Review by Regions

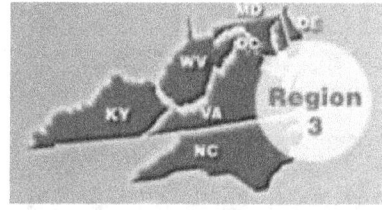

Region 3: District of Columbia, Delaware, Kentucky (partial), Maryland, North Carolina, Virginia, West Virginia

Reponses received from District of Columbia, Delaware, Kentucky, Maryland, North Carolina, and Virginia.

DISTRICT OF COLUMBIA

Washington D.C.'s response included the following table and project descriptions:

System Quality Attribute	Crash	Roadway	Driver	Vehicle	Citation/ Adjudication	Injury Surveillance
Timeliness	X^1				X^3	X^4
Accuracy	X^1	X^2			X^3	X^4
Completeness	X^1	X^2			X^3	X^4
Consistency	X^1	X^2				X^4
Integration						
Accessibility					X^3	

Notes:

1) In May 2008, the Metropolitan Police Department implemented an automated tool for field crash data collection. It is a Web-based, real-time system. Full implementation was expected by summer of 2010. Numeric performance measures included only qualitative measures.

2) The State lists three roadway-related projects: (a) master address repository, (b) GIS repository platform, and (c) Service-Oriented Architecture for Transportation Services (Transportation Services Modernization Program). Performance measures were not supplied.

3) The State lists four projects affecting citation/adjudication data: (a) Nightly data exchange between the court and DMV; (b) LEADRS effort to reduce DUI processing time; (c) consolidation of tickets and driver history information; (d) online viewing of citation information. Detailed project descriptions and performance measures were not provided.

4) The EMS electronic database/repository project is designed as a centralized EMS run report database. The project began in 2006 and was completed in fall 2008. One hundred percent of EMS run reports are input to the database within 10 days of the incident. As of June 2010, no additional detailed project description and performance measures for accuracy and consistency were provided.

Appendix B: State Review by Regions

DELAWARE

Delaware's response included the following table and a report from its performance measurement tracking system providing quantitative quality improvement metrics.

System Quality Attribute	Crash	Roadway	Driver	Vehicle	Citation/ Adjudication	Injury Surveillance
Timeliness	X^1				X^3	
Accuracy					X^3	
Completeness					X^3	
Consistency						
Integration					X^3	
Accessibility	X^2					

Notes:
1) The State listed numerous projects on the TRIPRS Web site. Two were highlighted in its response. The first deals with timeliness of TraCS crash transfers between the Delaware State Police (DSP) and the Delaware DOT. Timeliness went from 90+ days in 2007 to 30 days in 2008. No further information was provided about how they are progressing on the Stated goal of daily transfer between agencies.

2) The Office of Highway Safety, DSP, and local agencies were provided access to crash data via the CHAMPS analysis and reporting system. User tracking shows that 53 percent of local law enforcement agencies were using CHAMPS by the end of 2008. No further information was provided by the State to determine if the user population has grown since then and if access was expanded beyond the law enforcement community.

3) The Delaware Criminal Justice Information System completed deployment of an electronic citation module for law enforcement and the courts. CJIS also sends data electronically from the courts to the DMV. The program is designed to improve timeliness and completeness of information in CJIS and the DMV, increase accuracy using GPS coordinates, and improve the ability to link citation and crash data where applicable. According to performance measure data for 2008 and prior, timeliness of citation information improved from 7.5 days pre-2007 baseline to 1.7 days in 2008. Completeness, measured in terms of the number of electronic citations with GPS coordinates, has increased from zero in 2007 to 67,059 citations in 2008. The project review team added a check mark for accuracy (in addition to completeness noted by the State) as the addition of GPS coordinates will assist the State to obtain more accurate location information. It is possible that the State intended to use the presence of GPS coordinates in the citation record is an indication of its ability to link citation and crash data. No additional information was provided by the State on more recent data (2009 and beyond) and a method to measure data integration was not provided.

KENTUCKY

The State's response did not include a formal completion of the questionnaire. The following table is based on data submitted by the Kentucky Hospital Association only.

Quality Attribute \\ System	Crash	Roadway	Driver	Vehicle	Citation/ Adjudication	Injury Surveillance
Timeliness						
Accuracy						X[1]
Completeness						X[1]
Consistency						
Integration						
Accessibility						X[1]

Notes:

1) The Kentucky Hospital Association received funding in three consecutive years to record Emergency Department treatments and observation stays for an estimated 1,740,000 patients annually. The performance measures include total number of records submitted (completeness) and accuracy of the submissions based on the percentage of case records passing the edit checks imposed by KHA (current target is 99 percent error free submission). For the two complete years of reporting to date, the system has met its reporting level targets. No data were supplied on whether the accuracy targets are being met. No further information was provided by the State about obtaining accuracy metrics and current year (2010) data upon completion of the second quarter data submissions. The project's funding ended in June 2010, but no additional information from the State was provided about how the project will be maintained in the absence of grant funding.

MARYLAND

The State submitted a link to its 2009 Section 408 Grant Application as a source of project descriptions and progress reports. Its table was developed based on that report and follow-up e-mail with the Highway Safety Office's Traffic Records Coordinator.

System Quality Attribute	Crash	Roadway	Driver	Vehicle	Citation/ Adjudication	Injury Surveillance
Timeliness						
Accuracy					X^2	
Completeness	X^1				X^2	
Consistency						
Integration	X^3		X^3	X^3	$X^{2,3}$	X^3
Accessibility	X^3		X^3	X^3	X^3	X^3

Notes:

1) For CMV-involved crashes, the State implemented a project to improve completeness of driver and vehicle information. In 2008, measures of completeness rose to 79 percent for driver information (from a baseline of 72 percent) and 59 percent (from a baseline of 36 percent) for vehicle information.

2) Maryland is using an electronic citation system (E-TIX) that is capable of capturing GPS coordinates as part of the location information. Completeness of GPS reporting went from a baseline of 0.71 percent in the first half of 2008 to 13.6 percent through early 2009. The State did not provide data that is more recent and to clarify the calculation of the completeness metric. It appears that it is simply reporting the percentage of citations processed electronically, whether or not the GPS information is accurate. There is an inconsistency in the 2009 progress report between the narrative description of the metric and the specification of how the metric is calculated. One says the measure is based on the proportion of all citations with GPS coordinates reported. The other says that the measure is based on the proportion of citations received electronically. The two methods are equivalent only if there is a 1:1 correspondence between the presence of GPS coordinates and electronic reporting. That would imply that electronic records without GPS data are rejected or corrected. In addition, no mention was made of GPS accuracy.

3) The Maryland Motor Vehicle Administration is managing a motorcycle safety project in cooperation with the University of Maryland National Study Center for Trauma and EMS. This study is driver and vehicle centered, with linkage of those two datasets with crash, citation, and injury surveillance data. Statistics on linkage and the safety data analyses will be produced by the National Study Center in 2011. The project is ongoing.

NORTH CAROLINA

The State submitted the following table and associated project descriptions:

System / Quality Attribute	Crash	Roadway	Driver	Vehicle	Citation/ Adjudication	Injury Surveillance
Timeliness	X^1				X^3	X^5
Accuracy	X^1	X^2			X^3	X^5
Completeness	X^1	X^2			X^3	X^5
Consistency	X^1				X^3	X^5
Integration		X^2			X^3	X^5
Accessibility		X^2			$X^{3,4}$	X^5

Notes:

1) The electronic submission of crash reports is a TraCS implementation that began in 2005. The State Highway Patrol and over 80 local agencies participate, resulting in electronic submission of over 33 percent of all crash reports. Timeliness overall in the central crash repository has improved from over 90 days post-crash to fewer than 7 days. No quantitative measures of accuracy, completeness, and consistency were provided.

2) The Local Roads Data Enhancement and Dissemination project was designed to support electronic collection and sharing of local road data among NCDOT and other State and local agencies. No quantitative measures for the data quality improvements noted were provided.

3) The eCITATION project was implemented in December 2003. The State provided timeliness data for 2005 through January 2010 showing that as the percentage of electronic citations grew, there was a corresponding reduction in the number of days post-issuance that the citations are entered onto the statewide citation repository. Current data indicates timeliness is now better than 3 days post issuance, down from just under 8 days in 2005. No quantitative measures for the other data quality improvements noted were provided.

4) The North Carolina Warrant Repository (NCAWARE) is a Web-based system for magistrates and allows law enforcement to search statewide for outstanding warrants. Not all counties are using the system at present. No detailed information on project status and performance measures was provided.

5) The State submitted performance measures for Injury Surveillance data, but did not list specific projects. No information was provided on specific projects for EMS, trauma, and emergency department data. EMS and ED data are updated daily within 24 hours. Trauma data are updated weekly with 100 percent completeness. Performance measures other than timeliness and completeness were not quantified.

Appendix B: State Review by Regions

VIRGINIA

Virginia's response included the following table along with supplemental information.

System Quality Attribute	Crash	Roadway	Driver	Vehicle	Citation/ Adjudication	Injury Surveillance
Timeliness	X^1					X^3
Accuracy	X^1					X^3
Completeness	X^1					X^3
Consistency	X^1					X^3
Integration	X^2	X^2	X^2	X^2	X^2	X^2
Accessibility	X^1					X^4

Notes:

1) Virginia's response provided qualitative information regarding each of the Crash data quality attributes and improvements based on the implementation of the new Traffic Records Electronic Data System (TREDS) crash reporting system. Consistency, as measured by MMUCC compliance increased in 2007 from 55 percent to 80 percent because of a redesign of the crash report form. The State did not provide documentation that the other data quality improvements were demonstrated with performance measures. A pilot test of electronic field data collection and submission began in August 2009. A June 2010 follow-up with the State showed that the TREDS project appears to have valid measures of data quality improvement for timeliness, completeness, and integration, in addition to the consistency measure reported in the original submission (for MMUCC compliance). As of March 2011, the State has achieved approximately 50 percent electronic data collection and 20 percent electronic data transfer. The plan is for 95 percent electronic by the end of 2011.

2) A follow-up e-mail from the State also indicated that TREDS integrates with the driver and vehicle databases in the DMV. In addition, the State's CODES project is listed as having successfully integrated Crash and Injury Surveillance data. Performance measures were not provided. As of the 2011 Traffic Records Assessment, TREDS has been able to provide metrics and/or data analyses using merged data from crash, roadway, driver, vehicle, citation and injury surveillance.

3) The Virginia Department of Health/Office of Emergency Medical Services has developed a Web-based reporting system that is NEMSIS-compliant for State-level data reporting. Individual EMS providers are given access to their own data through the Web portal. The system is described as improving timeliness, completeness, accuracy, and consistency, but no metrics or performance measures were supplied or described. The project was in the pilot stage as of early 2010. All of the performance improvements described for the EMS run reporting system are qualitative, not quantitative. It is likely that at least

consistency will have an actual measure as the State says the data definitions were based on the NEMSIS standard. Thus, it is assumed that they can say what level of NEMSIS compliance they have achieved. All of the other indicators of improvement in the follow-up report are limited to qualitative statements.

4) The Virginia Codes project Web site (www.vacodes.com) provides access to query and reporting functions. The Web site is open to all users without a registration requirement. Information about the use of the Web site was not provided, but may form the basis of a metric for accessibility. The Virginia CODES project has measures of successful data integration of crash and injury surveillance data sources including EMS, trauma, hospital discharge, and vital records. The data was not supplied, but CODES project managers can readily provide percent match and the number of years of linked data.

Appendix B: State Review by Regions

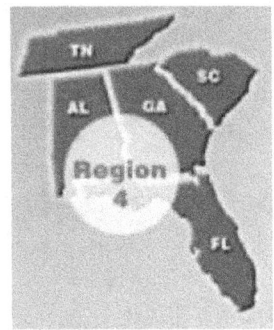

Region 4: Alabama, Florida, Georgia, South Carolina, Tennessee

Responses were received from all States except South Carolina.

ALABAMA

The State submitted the following table along with supplemental information:

System Quality Attribute	Crash	Roadway	Driver	Vehicle	Citation/ Adjudication	Injury Surveillance
Timeliness	X^1	X^2	X^3	X^4	X^5	X^6
Accuracy	X^1	X^2	X^3		X^5	X^6
Completeness	X^1				X^5	X^6
Consistency	X^1				X^5	X^6
Integration	X^1	X^2	X^3		X^5	$X^{6,7}$
Accessibility	X^1	X^2	X^3	X^4	X^5	X^6

Notes:

1) The State implemented an electronic crash reporting system beginning in July 2007 with final rollout in June 2009. Performance measures were supplied for timeliness, accuracy, completeness, and MMUCC compliance (now 100 percent). Qualitative descriptions of improved internal consistency, integration, and accessibility were provided without quantification. Timeliness of crash data is projected to improve from 40 days down to about 3 days due to use of electronic field data collection and electronic data transfer. For other data quality performance measures of crash data, the information is qualitative only. They did not collect pre-measures of data quality. The same is true for roadway information

2) A series of projects managed by the Alabama DOT (ALDOT) implemented a GIS-based location and analysis system. The system includes roadway inventory and crash data so that the State has improved the ability to conduct safety analyses based on roadway features and specific spot studies. Performance measures were not supplied to determine which of the project benefits could be quantified.

3) The Alabama Law Enforcement Tactical System (LETS) and the associated Centralized Agency Management System (CAMS) are described as automation efforts designed to increase data accessibility, timeliness, integration, and

accuracy. Performance measures were not supplied, but the supplied qualitative descriptions of improvement indicate that some metrics may have been collected for at least some of the data quality attributes.

4) A brief description of a project providing field officers with vehicle information was provided--a portable version of the LETS software called LETSGo. No further information was received to determine if any performance measures were collected regarding timeliness and accessibility of the data.

5) The Alabama electronic citation program (eCite) provides for field data collection and electronic submission of citation data to the courts. The eCite software includes error checking to improve both accuracy and completeness (both Stated to be near 100 percent) for electronic citations. Qualitative descriptions of improved integration and accessibility based on e-Cite implementation also was provided. The State is estimating a large revenue increase based on increased citation issuance from the electronic citation project (eCITE). The estimates are as high as $50 million increase, but the number should not be considered final. This is indicated as a green color for completeness.

6) The State implemented a NEMSIS-compliant EMS run reporting system with an electronic patient care report (ePCR). No further information was provided on year-end status reports for 2009 because many of the projects milestones were scheduled for completion by December. Performance measures for timeliness, accuracy, completeness, and consistency with the NEMSIS standard were described. Qualitative measures of internal consistency, integration, and accessibility were described. 2009 year end measures were not provided and it was not possible to determine if any of the qualitative measures can be quantified.

7) The Alabama CODES project produced integrated crash and injury surveillance data. The ability to link crash and EMS run report data was quantified in a performance measure. Linkage of other datasets was not quantified in the State's submission, but medical outcome data was mentioned as being linked as well.

Appendix B: State Review by Regions

FLORIDA

The following table and supplemental information were provided by the State:

System Quality Attribute	Crash	Roadway	Driver	Vehicle	Citation/ Adjudication	Injury Surveillance
Timeliness	X^1	X^2		X^3	X^4	X^5
Accuracy	X^1	X^2		X^3	X^4	X^5
Completeness	X^1	X^2		X^3		X^5
Consistency	X^1	X^2		X^3		X^5
Integration	X^1	X^2		X^3		X^5
Accessibility	X^1	X^2		X^3		X^5

Notes:

1) Florida referred to several projects listed in the TRIPRS Web site to provide support for the check marks for the crash system data quality improvements. These are FL-03, FL-06, FL-07, FL-08, FL-10, and FL-16. FL-03 is a training project (a workshop for trainers). FL-06 is a project to support electronic submission of crash data from local agencies. FL-07 is a project to establish an integrated data repository of Traffic Records Information. FL-08 is a project to support the State's implementation of TraCS. FL-10 is a project to support electronic submission of crash reports by the Florida Highway Patrol (FHP). FL-16 is a project to support (generally smaller) agencies that do not have electronic crash reporting capabilities by implementing a Web-based reporting tool. A new project (not listed in TRIPRS) was provided as part of the State's response. MMUCC and SAFETYNET compliance are improved measurably with the new crash report form. Data on exact measures of consistency with the standards is pending, but anticipated to be over 97 percent for MMUCC and 100 percent for SAFETYNET.

   The State moved to the new form on January 1, 2011. DHSMV is not accepting electronic data from any agency other than FHP now, but there are several agencies using TraCS that anticipate sending data electronically once the State's TraCS implementation is updated to the new form and is certified as meeting the data submission standards. Green indicators for crash accuracy and integration are based on the 98 percent success rate achieved for mapping crashes using the statewide base map. Crash reporting performance measures reported on TRIPRS are limited to timeliness and a completeness measure that really measures integration of crash and roadway data. No further information was received about general performance measures and additional project-specific measures for the past five years.

2) Florida referred to FL-04 and FL-05 project descriptions on the TRIPRS site as backup for the roadway data quality improvements listed. FL-04 is a project to collect roadway characteristics (inventory data) for off-system (I.e., not State-

maintained) roadways and combine the data with crash data that are based on verified location information. FL-05 is a project to establish a unified base map (GIS) for all roadways in Florida. The TRIPRS site has one performance measure showing the number of counties for which Florida DOT has a usable base map containing all public roadways, but the data presented appear to be in error. The two roadway projects are not sufficient to justify the checkmarks under roadway data timeliness and accessibility improvement, but with the addition of the data repository created under FL-07, those two improvement indications may be justified. Green indicators for accessibility of roadway and Injury Surveillance data are based on Web-based secure access systems with performance measures as listed on TRIPRS of number of downloads.

3) The Department of Highway Safety and Motor Vehicles (DHSMV) implemented a project to record temporary tag sales/registrations) at automobile dealers electronically. The project has clear implications for timeliness and accessibility of this subset of registrations (temporary tags issued at the time of purchase from a dealership). Performance measures were not supplied, but one for timeliness should be available or easy to compute since performance went from manual delayed entry to real-time entry and accessibility. It is not clear why the other data quality attributes under vehicle data were also checked. No additional information was provided.

4) Florida referred to projects FL-08 and FL-17 on the TRIPRS site as leading to improvements in citation/adjudication data quality. FL-08 is the TraCS implementation support project, which included an e-citation component. FL-17 is a project specifically designed to improve timeliness and accuracy of Universal Traffic Citation data by analyzing the causes of delay and inaccuracy of data for DUI and serious traffic violation cases.

5) Florida referred to projects FL-02, FL-07, and FL-09 on the TRIPRS site as leading to data quality improvements in injury surveillance systems. FL-02 and FL-09 are implementations of field data collection and centralized reporting of EMS run report data, respectively. FL-07 is the centralized data repository project that includes integration of EMS and other traffic records data sets. No information was provided about general and project-specific performance measures.

Appendix B: State Review by Regions

GEORGIA

Georgia submitted the following table and supporting information:

System Quality Attribute	Crash	Roadway	Driver	Vehicle	Citation/ Adjudication	Injury Surveillance
Timeliness	X^1				X^3	
Accuracy	X^1		X^2		X^3	
Completeness					X^3	
Consistency						
Integration						
Accessibility	X^1		X^2		X^3	X^4

Notes

1) Georgia is in the process of implementing an Indiana-style crash management contract in which the chosen vendor manages the centralized collection, storage, and accessibility of crash records. The plan is for 75 percent electronic submission of data (up from current 7 percent) by October 1, 2010. Follow-up with the State at the end of the Federal fiscal year is warranted to see how the project is progressing, and to obtain measures of performance. No information about performance measures on current data was provided. The most recent data for crash data quality on the TRIPRS site appear to be from 2007, except for data on fatal crashes.

2) Driver Services has revised the driver's license to include security and machine-readable features. These changes should help law enforcement auto-populate fields on crash report and citation forms. Performance measures related to this project were not apparent in the State's submission or the TRIPRS site.

3) The creation of a centralized citation data warehouse and support for electronic transmission of data from the courts has improved timeliness of reporting conviction data to Driver Services. Performance measures on TRIPRS (GA-PM-03) indicate that the days from citation issuance to transmission of a conviction to Driver Services have dropped from 180 days (pre 2007) to 67 days (2009).

4) The Online Analytical and Statistical Information System (OASIS) has been expanded to allow public access to emergency department data for persons injured in motor vehicle crashes. Accessibility has been measured by the number of downloads of data from the site. No data were provided to use as a baseline and to see if information is available on the types of people using the site (i.e., is there increased volume and variety of uses, or has the site seen increased use mainly from previous customers).

Appendix B: State Review by Regions

TENNESSEE

Tennessee submitted the following table along with supplementary information:

System Quality Attribute	Crash	Roadway	Driver	Vehicle	Citation/ Adjudication	Injury Surveillance
Timeliness	X^1					
Accuracy	X^1					
Completeness						X^3
Consistency	X^1					
Integration						X^3
Accessibility	X^1				X^2	

Notes:

1) Tennessee provided the following performance measures for crash data quality:
   Regarding **Crash Timeliness**, Tennessee has demonstrated improvements with:
   - An overall decrease from 80 percent in 2007 to <1 percent in 2009 in the percentage of crash reports that are entered into the database more than 90 days after the crash event for 95 percent of the data;
   - An increase from 39 percent in 2007 to 51 percent in 2009 in the overall percentage of crashes entered into the statewide database within 30 days of the crash event for 95 percent of the data; and
   - An increase from 2 percent in 2007 to 91 percent in 2009 in the overall percentage of crashes entered into the statewide database less than 60 days but more than 30 days of receipt of the crash event.

   Regarding **Crash Accuracy**, Tennessee demonstrated improvements with:
   - An increase from 62 percent in 2007 to 100 percent in 2009 of motor carriers in the crash database that are matched in the MCMIS.

   Regarding **Crash Consistency**, Tennessee demonstrated improvements with:
   - An increase from 25 percent in 2006 to 95 percent in 2009, overall, of MMUCC-required data elements present in the crash database.

   Regarding **Crash Accessibility**, Tennessee demonstrated improvements with:
   - An increase from 5 percent in 2007 to 19 percent in 2009 in the percentage of local enforcement agencies using online retrieval of crash and statistical reports. Specific projects producing these results are described on the TRIPRS site under TN-P11 (TRCC administration); TN-P22 (electronic upload of crash data for the Tennessee Highway Patrol); TN-P23 (data entry of crash data); TN-P61 (Tennessee Integrated Traffic Analysis System), and TN-P62 (Development of Standardized Reports From Crash Data).

2) Tennessee provided the following performance measures for citation/adjudication data quality:

Regarding **Citation/Adjudication Accessibility**, Tennessee has demonstrated improvements in:
- An overall increase from 2752 in 2007 to 5176 in 2009 in the number of users subscribed to and using the criminal justice portal; and
- An increase from 2 in 2007 to 8 in 2009 in the number of traffic courts that use the portal to access driver history while processing driver cases.

The submission lists project TN-P41 (Integrated Criminal Justice Portal) as relevant to this improvement.

3) Tennessee provided the following performance measures for injury surveillance data quality:

Regarding **Injury/Surveillance Completeness**, Tennessee has demonstrated improvements in:
- An increase from 90 in 2006 to 189 in 2009 in the total number of ambulance services submitting EMS run reports.

Regarding **Injury/Surveillance Integration**, Tennessee has demonstrated improvements in:
- A 100 percent participation of the total number of trauma centers submitting patient data to EMITS.

The submission lists project TN-P52 (Implementation of EMITS and trauma registry databases) as relevant to the measured improvements.

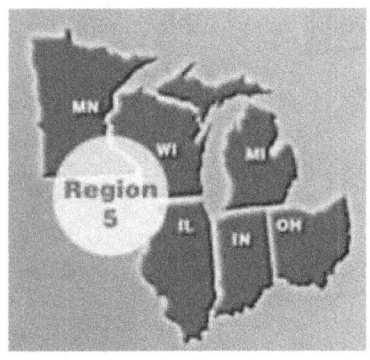

Region 5: Illinois, Indiana, Michigan, Minnesota, Ohio, Wisconsin

Responses were received from Illinois, Indiana, Michigan, and Minnesota.

ILLINOIS

Illinois provided project descriptions for two information technology projects managed by the DOT. The following table was created based on these descriptions.

System Quality Attribute	Crash	Roadway	Driver	Vehicle	Citation/ Adjudication	Injury Surveillance
Timeliness	X^1					
Accuracy	X^1					
Completeness						
Consistency						
Integration						
Accessibility	X^2					

Notes:

1) The Crash Information System (CIS) is the central crash repository and associated data entry processes managed by the DOT. Performance measures defined for the project include two measuring timeliness (average days from crash even to data entry, and average days from data entry until the data are available for analysis), and one measuring accuracy (percentage of records containing no more than a specified number of critical errors). Performance measures were supplied as follows:
CIS:

a. Average number of days from the date of a reported crash until it is entered into the State crash file (using the 95 percent of cases).
   1. Average # of days for the crash to be reported to IDOT is 15 days (paper – 20.32 days, MCR – 7.06 days)
   2. Average # of days for IDOT to do the data entry is 25.63 days (paper – 37.70 days, MCR – 7.60 days)
   3. Average # of days for IDOT to do the location entry is 24.45 days (paper – 24.47days, MCR – 24.43 days)

Appendix B: State Review by Regions

b. Average number of days from the data entry date until it is available for analysis (using the 95 percent of cases).
24.45 days (paper – 24.47days, MCR – 24.43 days)

c. Percentage of records on the State crash file that contain no more than a specified number of critical data elements with errors.

d. Percentage of time unknown values are used in the critical data elements is 9.81 percent (paper – 12.33 percent, MCR – 6.05 percent)

2) The Safety Data Mart is designed to support users' needs for crash records access. Performance is measured in terms of total number of agencies and users accessing the crash files

a. Safety Data Mart (data accessibility). The performance measures listed for this project are not reported on TRIPRS (although, there is a measure of the number of reports available).

b. Number of authorized agencies capable of accessing crash files.
Fewer than 10 agencies have access to our database

c. Number of users accessing crash files

SafetyDataMart: Executive Dashboard	April 23, 2010 - April 30, 2010

Traffic Analysis	
Total visits:	212
Total page views:	10,956
Total hits:	18,730
Total megabytes transferred:	0
Average visits per day:	28
Average visits per week:	197
Average visits per month:	Insufficient data
Average pages viewed per visit:	52
Average pages viewed per day:	1,455
Highest volume time of day:	2 p.m. - 3 p.m.
Highest volume day of the week:	Friday
Highest volume month:	April 2010

INDIANA

Indiana submitted a May 2009 report titled *State of Indiana Traffic Records: The Evolution* in response to the request for projects involving traffic records improvement. The report deals exclusively with Crash reporting in the State – the main emphasis area since the 2005 NHTSA Traffic Records Assessment. The following table was developed based on the information provided in the 2009 report:

System Quality Attribute	Crash	Roadway	Driver	Vehicle	Citation/ Adjudication	Injury Surveillance
Timeliness	X[1]					
Accuracy	X[1]					
Completeness	X[1]					
Consistency						
Integration						
Accessibility						

Notes:

1) The State has privatized its crash data collection and reporting under a contract with a vendor who is now responsible for meeting data quality targets and providing revenue to the State for sales of crash report copies. The May 2009 report includes performance measures of the percentage of crash reports submitted electronically, the timeliness of those reports, and the percentage of reports rejected (i.e., failing accuracy or completeness checks).

Comparison of Electronic Submissions vs. Rejected Reports

Year	# Paper	# eVCRS	Total	Rejected	% Paper	Error Rate %
2003	212,784	789	213,573	79,952	38%	37.0%
2004	171,009	38,291	209,300	64,940	38%	31.0%
2005	140,525	68,573	209,098	55,285	39%	26.0%
2006	81,106	113,600	194,706	26,753	33%	14.0%
2007	15,762	129,624	145,386	4,953	31%	3.0%
2008	3,945	201,468	205,413	60	2%	1.5%

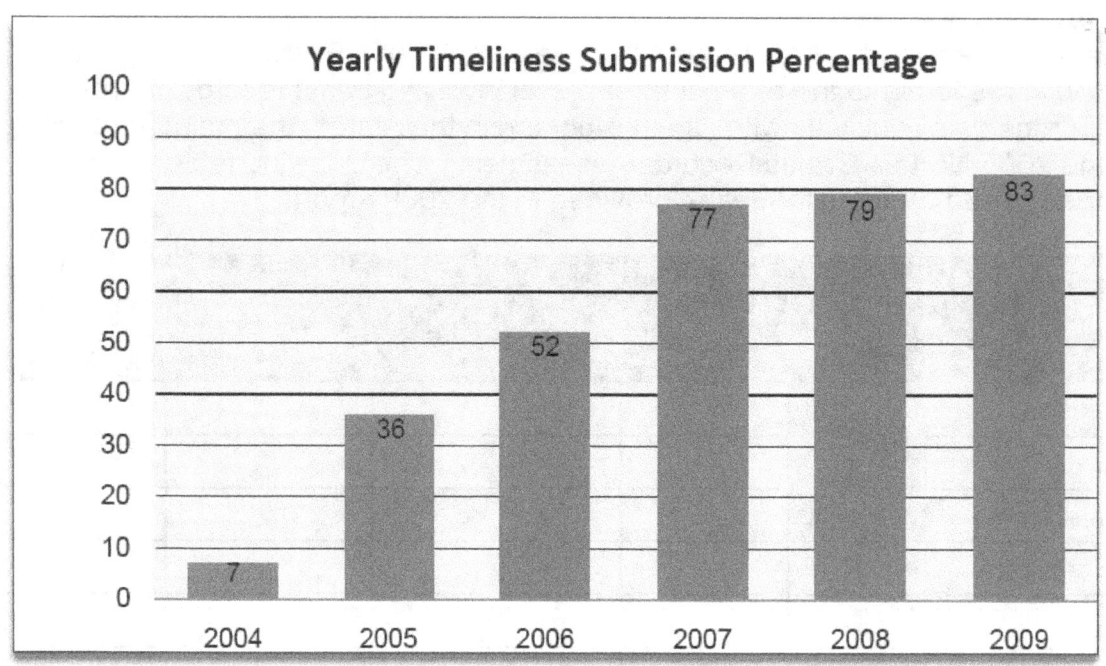

In a follow-up, the State provided up-to-date data on the crash reporting system. Indiana is continuing to make updates to the eVCRS and ARIES software that will improve, for example, coding of crash locations (using a point-and-click map interface), and requiring a specific answer in key fields (prompting the officer if the code for "other" is used in those fields). The green indicators in the performance measures for timeliness, accuracy, and completeness are based as much on the original submission as on the follow-up, which mainly provided background information. The original data was already sufficient to document the system improvements. The following provide performance measures data:

MICHIGAN

Michigan completed Part 2 of the questionnaire, providing project-level descriptions of significant traffic records improvements. The following table is based on those project descriptions.

Quality Attribute \\ System	Crash	Roadway	Driver	Vehicle	Citation/ Adjudication	Injury Surveillance
Timeliness	X^1				X^2	
Accuracy	X^1					
Completeness	X^1					
Consistency						
Integration					X^2	
Accessibility	X^1				X^2	

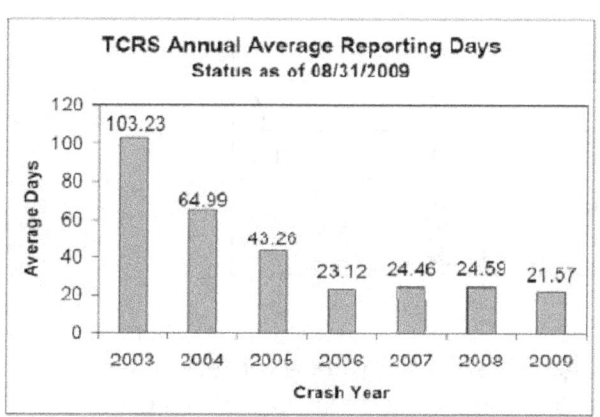

TCRS Annual Average Reporting Days
Status as of 08/31/2009

Notes:

1) Beginning in 2002, the Crash Process Redesign has followed a phased approach to the release of upgraded software and procedures (now in Release 8). The Traffic Crash Reporting System (TCRS) and related projects have yielded major improvements in timeliness, accuracy, and completeness (as shown in the performance measures provided in the submission). There are also benefits in data accessibility and analysis, but these may not have been measured using a data quality metric.

2) The Judicial Data Warehouse project began in 2002 with trial court information from 13 of 15 counties in Michigan's Upper Peninsula. In 2006 and 2007, the project was expanded to include the majority of courts in the State. As of 2008, 81 of 83 counties, and 219 of 255 trial courts are participating in the data warehouse. The State provided the 2010 Section 408 progress report for the Judicial Data Warehouse project. The project has succeeded in making data available to users within hours (less than a day in most courts). The year 2010 saw more courts join the project.

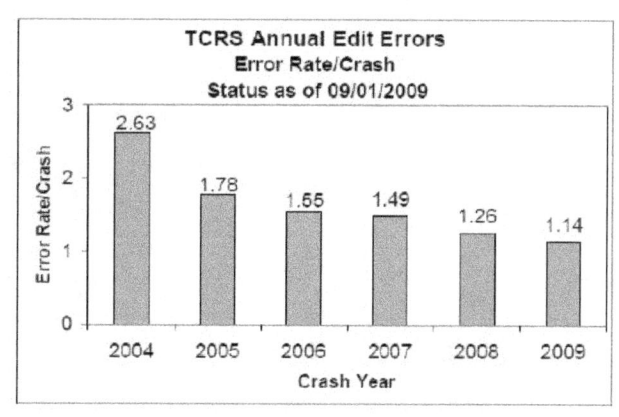

TCRS Annual Edit Errors
Error Rate/Crash
Status as of 09/01/2009

Appendix B: State Review by Regions

Also provided multiple reports of crash data quality metrics – too numerous to reproduce here, but measures include timeliness (multiple measures), accuracy/completeness (multiple measures), and consistency. This has resulted in reduced delays in data accessibility as shown in the following table:

County Courts		Number of Hours	
		2008	2009
Bay County		70	17
Allegan County		49	42
Hillsdale County		47	11
Isabella County		36	15
Charlevoix County		10	5
Barry County		47	7
Marquette County		48	9
Ottawa County		40	7
Wayne County		118	14
Average Time for n (hours)		52	14

In addition, the project is designed to provide better integration of citation and crash information. This could be determined by State follow-up to quantify progress on this goal.

MINNESOTA

Minnesota provided a project description in response to part 2 of the questionnaire. The following table is based on the project description:

System Quality Attribute	Crash	Roadway	Driver	Vehicle	Citation/ Adjudication	Injury Surveillance
Timeliness	X[1]					
Accuracy	X[1]					
Completeness	X[1]					
Consistency	X[1]					
Integration	X[1]					
Accessibility	X[1]					

Notes:

1) The Crash Database Interface for Law Enforcement Agencies allows crash data to be entered once into an agency's records management system and passed electronically to the central crash repository at the Department of Public Safety, eliminating duplicate data entry. The project includes implementation of error checking for completeness, accuracy, and internal consistency. Performance measures for timeliness and completeness. The State also provided information on integration between local and State records, which also show improvement in records accessibility. Follow-up is required to determine whether use of the interface has expanded beyond the Minnesota State Patrol (MSP) or on what timeline that expansion will take place.

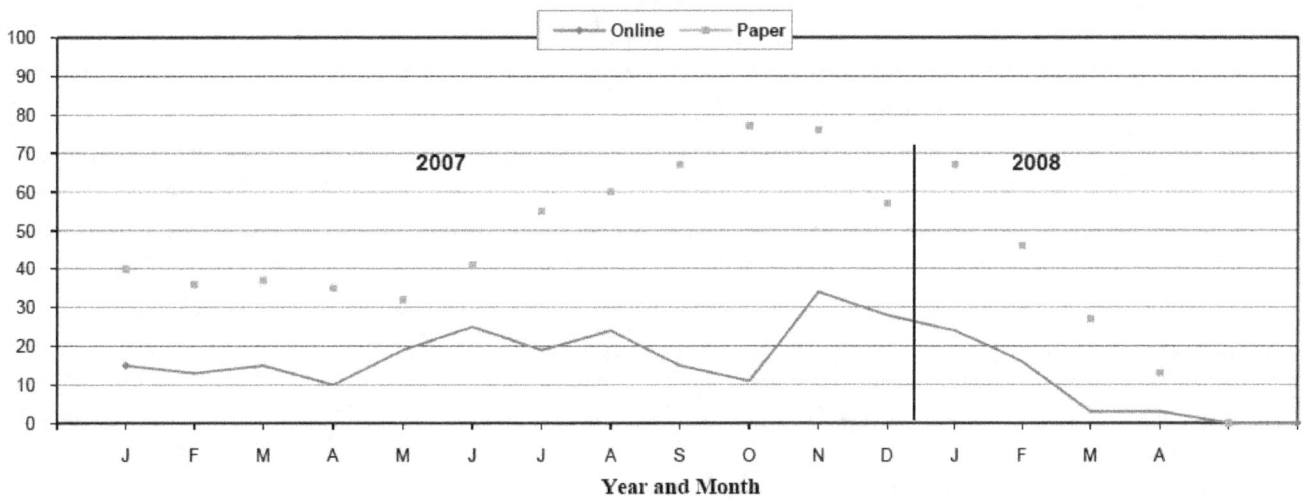

Timeliness Performance
Measure: Average Days from Crash to DVS Entry Datee
By Reporting Method

Appendix B: State Review by Regions

The following seven data elements showed a statistically significant reduction in the number of times they were left blank on PARs submitted by the MSP during the months of October through December.

1: Most Harmful Event:
Baseline: (Oct-Dec 2007)   1.24%
Current: (Oct-Dec 2008)   0.82%

2: First Harmful Event:
Baseline: (Oct-Dec 2007)   1.11%
Current: (Oct-Dec 2008)   0.87%

3: Driver License Status:
Baseline: (Oct-Dec 2007)   0.65%
Current: (Oct-Dec 2008)   0.47%

4: Ejected from Vehicle:
Baseline: (Oct-Dec 2007)   6.90%
Current: (Oct-Dec 2008)   3.30%

5: Airbag Deployment:
Baseline: (Oct-Dec 2007)   6.70%
Current: (Oct-Dec 2008)   3.40%

6: Safety Equipment Use:
Baseline: (Oct-Dec 2007)   6.20%
Current: (Oct-Dec 2008)   2.80%

7: Safety Equipment Type:
Baseline: (Oct-Dec 2007)   6.20%
Current: (Oct-Dec 2008)   2.80%

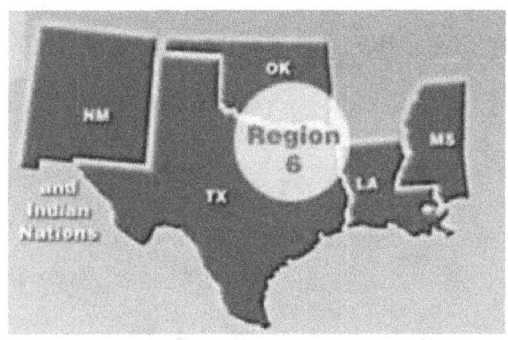

Region 6: Louisiana, Mississippi, New Mexico, Oklahoma, Texas

Responses were received from Louisiana and Texas.

LOUISIANA

Louisiana provided three project descriptions related to improvements in the crash data system. The following table is based on a combination of the three projects.

System Quality Attribute	Crash	Roadway	Driver	Vehicle	Citation/ Adjudication	Injury Surveillance
Timeliness	X^1					
Accuracy	X^2					
Completeness						
Consistency						
Integration						
Accessibility	X^3					

Notes:

1) The Highway Safety Research Group at Louisiana State University (LSU) manages the LACRASH Web site as a service to the State, and provides data entry/management, equipment, and field data collection software in support of crash reporting by law enforcement agencies. Efforts to improve timeliness include promotion of field data collection (electronic reporting has increased from 32,000 reports in 2005 to over 60,000 electronic reports in 2008), advanced pen-based and scanner solutions for agency-level data capture, and online publishing of statistics for use by law enforcement agencies. Timeliness measures for each reporting agency are published on the LACRASH Web site and help to bring focus to the issue of timely reporting. In June 2010, LSU provided updated timeliness measures (including calendar 2009 data). The State is in the process of calculating metrics of accuracy and accessibility. An accuracy measure is being calculated based on a CDIP-sponsored audit of crash reports. The accessibility measure is based on a survey of data users and may be less quantifiable than accuracy. The figure below shows the progress made over the past several years.

Appendix B: State Review by Regions

Year	% < 30	% < 60	% < 90	Avg # Days
2006	39	41	44	118
2007	46	51	62	101
2008	63	82	82	78

In 2008, paper reports averaged 96 days while electronic reports averaged 22 days.

2) Accuracy improvements are expected based on a targeted audit process being implemented by LSU in cooperation with the law enforcement agencies. The process involves focusing on critical data elements (such as location) and selection of a random sample of crash reports from each law enforcement agency. The crash reports are examined for errors and a report is provided to the submitting agency. Follow-up is required as this project was in the early stages of implementation at the time of the response.

3) The LACRASH Web site includes utilities for users to query, filter, and generate summary reports of crash data. A new mapping function is slated for implementation in early 2010. The Web site also reports performance measures of quality for the crash reporting system as a whole, and by agency. Follow-up is required to obtain performance measures for data accessibility and use of the system's analytic tools.

Appendix B: State Review by Regions

TEXAS

Texas provided the following table and supporting project description:

System Quality Attribute	Crash	Roadway	Driver	Vehicle	Citation/ Adjudication	Injury Surveillance
Timeliness	X[1]					
Accuracy	X[1]					
Completeness	X[1]					
Consistency	X[1]					
Integration						
Accessibility	X[1]					

Notes:

1) The Crash Records Information System (CRIS) has replaced the manual processing of approximately 600,000 crash reports submitted in Texas each year. The CRIS project included elimination of a 5-year backlog of crash data entry, support for analysis (response time to requests has dropped from 3 to 6 weeks down to 3 days), improved accuracy via a two-step entry and audit process with automated edit checks – including checks for incomplete data, improved internal consistency through training and error checking, and creation of an online portal to facilitate access to data and analytic reports. Performance measures were supplied for all of the data quality areas except completeness, which is now pegged at 100 percent (incomplete reports are rejected).

Appendix B: State Review by Regions

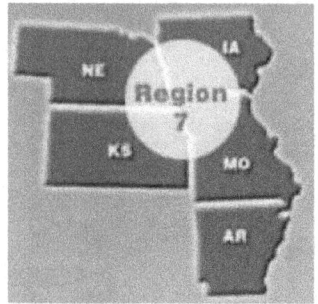

Region 7: Arkansas, Iowa, Kansas, Missouri, Nebraska

All five States in Region 7 responded to the questionnaire.

ARKANSAS

Arkansas provided the following table, and referred to the TRIPRS Web site for project descriptions and performance measures. Some check marks in the table were changed based on the information on TRIPRS.

System Quality Attribute	Crash	Roadway	Driver	Vehicle	Citation/ Adjudication	Injury Surveillance
Timeliness	X[1,2]				X[4]	X[5]
Accuracy	X[1]					
Completeness	X[1]					
Consistency	X[1,3]					X[5]
Integration	X[1]					
Accessibility	X[1]				X[4]	

Notes:

1) Arkansas is implementing TraCS. A detailed plan is required to determine the status of the roll out to the Arkansas State Police (ASP) and to local agencies. The performance measures on TRIPRS related to this project address timeliness and percentage of data received electronically from ASP only. Accuracy of location information is being improved through development of a point-and-click map interface for officers in the field.

2) The backlog has been reduced from 12 months to approximately 3 months based on the efforts of a vendor hired to enter the older reports.

3) MMUCC Compliance has increased to 73 percent because of a project to revise the crash report. The project also included training for law enforcement officers, but there is no specific performance measure that would quantify the accuracy benefits expected.

4) Timeliness of dispositions submitted by four municipal courts involved in a pilot study has improved from less than 15 percent entered within 40 days to over 99 percent entered within that target timeframe. In a related project, the number of courts updating the driver file daily has increased from zero in 2005 to five in the first half of 2009.

Appendix B: State Review by Regions

The EMS Data Injury Surveillance Continuation Project has helped the EMS reporting system reach the NEMSIS "Silver" standard of compliance. Performance measures tracking the number of EMS providers submitting data show that the system is meeting and exceeding its targets for deployment.

IOWA

Iowa submitted the following table and supplied project descriptions as indicated. The numbering of notes has been modified from the original submission.

System Quality Attribute	Crash	Roadway	Driver	Vehicle	Citation/ Adjudication	Injury Surveillance
Timeliness	X^1					
Accuracy	$X^{1,2}$					
Completeness	$X^{1,2}$				X^9	X^{10}
Consistency						X^{10}
Integration	$X^{3,4,5}$	$X^{3,4,5}$	$X^{7,8}$	X^7		X^8
Accessibility	$X^{3,4,5,6}$	$X^{3,4,5}$	X^8			$X^{8,11}$

Notes:

1) Project 1.1 in the submission is the statewide TraCS implementation. Performance measures are as follows.

Baseline year	2005	60% submitted electronically
	2008	81.4%
	2009	82% (preliminary estimate)

Timeliness of Reports using the "within 30 days" criterion and a comparison period of October to December each year.

Baseline year	2006	90.0%
	2007	92.9%
	2008	95.5%

2) Project 1.2 in the submission is designed to improve the collection of crash location data through use of a point-and-click map interface in the field data collection. The performance measure is percentage of crash reports received with latitude and longitude coordinates based on incident location tool use in the field. The baseline in 2005 was 50 percent of reports received with latitude and longitude coordinates. As of November 2009, 99 percent of electronic reports received included latitude and longitude coordinates.

3) The Geographic Information Management System (GIMS) is completing user testing in early 2010. The system is designed to improve the accuracy of GIS information for roadway, crash, and other geo-located information. Performance measures were not supplied.

4) The Linear Referencing System is designed to support easy integration of roadway inventory, crash, pavement management, bridge, and other databases. Location information in the system is accurate to better than five meters. No performance measures were supplied.

Appendix B: State Review by Regions

5) Project 6.1 in the submission is the Iowa Traffic Safety Data Service. This service is designed to provide analytic and data extract support to a wide range of users. Tracking of usage statistics show that the ITSDS has addressed approximately 600 requests from nearly 170 groups and individuals.

6) Project 6.3 in the submission is Analysis Tools Training. Training in use of available tools (in particular CMAT) for crash data analysis is provided to a wide variety of users, primarily drawn from State and local government. The user community for CMAT is nearing 600 people. Follow-up is required to obtain baseline information and the timeframe during which this project has operated.

7) Project 3.3 in the submission is Driver and Vehicle File Integration. A redesign of both the Driver System and integration with a redesigned Vehicle Registration and Titling System was completed in 2007. Performance measures were not supplied.

8) Project 3.1 in the submission is Driver Behavior Data Integration. This project involved data sharing to support epidemiological analysis and access to merged data for decision-making. While there are no formal performance measures for this effort, it has demonstrated successes in integrating data and making the data available for analysis.

9) Project 4.1 in the submission is the implementation and promotion of electronic citations using the TraCS software. The following performance measures were supplied.

Baseline year	2005	25.7%	electronically submitted
	2007	46.5%	electronically submitted
	2008	53.3%	electronically submitted

10) Project 5.1 in the submission is Improving EMS Data Collection--NEMSIS progress. This project is designed to assist EMS providers and their various software vendors in complying with the requirement to submit run report data to the Iowa EMS Patient Data Warehouse. In addition, the EMS data dictionary now exceeds the NEMSIS Silver standard. The following performance measure for run report submissions was provided.

Baseline year	2005	120,461 reports (50 % of call volume)
	2006	146,342 reports (61 % of call volume)
	2007	166,112 reports (69 % of call volume)
	2008	193,818 reports (81 % of call volume)

11) Project 5.2 in the submission is the CODES project. This project links EMS, other injury data, and crash data to support analysis of crash outcomes. Iowa CODES now has 11 years of linked data available for use. No performance measures were supplied.

KANSAS

Kansas provided the following table and associated project descriptions:

System Quality Attribute	Crash	Roadway	Driver	Vehicle	Citation/ Adjudication	Injury Surveillance
Timeliness	X[1,3]					X[5,6]
Accuracy	X[1,3]					X[5,6]
Completeness	X[1,3]					X[5,6]
Consistency	X[1]					X[5,6]
Integration	X[1,2,4]					X[5,6]
Accessibility	X[1,2]					X[5,6]

Notes:

1) The KKARS crash repository has been updated based on a redesigned crash report to increase MMUCC compliance. The system is now able to accept data electronically using an XML schema, which includes additional validation checks. Performance measures in TRIPRS show the following:

   - Crash timeliness (measured as percentage of reports processed within a 60-day target) has actually gotten worse, down to 18 percent processed on time in 2008.

   - MMUCC compliance is at about 50 percent.

   - FARS BAC reporting reached 96 percent in 2008.

   - Percentage of agencies submitting crashes electronically was 10 percent in 2009, down from 11 percent in 2008. The percentage of overall reports submitted electronically went up from 11 percent to 19 percent in the same period.

2) The Traffic Records System is a virtual data warehouse established in 2008. The initial datasets included crash and citation/adjudication.

3) The Kansas Law Enforcement Reporting System is intended to provide field data collection for a complete suite of law enforcement reports (including crash, incidents, and electronic citations).

4) Kansas is in the process of developing an e-citation application. The status and project description were incomplete in the State's submission and no details were available on TRIPRS. It appears this project is in its early stages (prototype under development).

5) The Kansas Emergency Medical Information System (KEMIS) project includes a new central repository and tools for field data collection for EMS run reports. The central repository has been implemented and the field data collection software has been implemented in 40 agencies. NEMSIS compliance has improved to include all 83 of the recommended State elements, included in 192 data elements overall.

6) The Kansas Trauma Registry is a statewide repository for trauma data submitted by hospitals.

Appendix B: State Review by Regions

MISSOURI

Missouri submitted the following table and associated project descriptions:

Quality Attribute \ System	Crash	Roadway	Driver	Vehicle	Citation/ Adjudication	Injury Surveillance
Timeliness	$X^{2,3,6}$		X^9		$X^{10,12}$	X^{13}
Accuracy	$X^{1,2,7,8}$		X^9		$X^{10,12}$	X^{13}
Completeness	$X^{2,3}$				$X^{10,12}$	
Consistency	X^8					
Integration	X^1					X^{13}
Accessibility	$X^{4,5}$		X^9		$X^{9,11}$	X^{13}

Notes:

1) The GPS Line Work Base Map project will expedite creation of a base map for all public roads. Standardized location coding will facilitate integration of crash and roadway data. Follow-up is needed to obtain values for the percentage of crashes that "land" automatically on the GPS line work.

2) The Law Enforcement Traffic Software (LETS) is used in the St. Louis area. This project is an improvement to existing software to support electronic transfer of crash data to the statewide STARS system. Performance measures of the electronic transfer are needed.

3) The Local Crash Report Electronic Transfer is designed to support electronic transfer of crash data for users other than LETS and the Missouri State Highway Patrol (MSHP). Performance measures of electronic transfer are needed.

4) The project to enhance the Statewide Traffic Accident Records System (STARS) system to provide Web-based TRACE Reports is a first step in making it easier for users to access standard and ad hoc reporting capabilities for crash data. Performance data on the time from report request to completion are required.

5) The STARS Web-based Statistical Reports project builds on the success of the Web-enabled TRACE reports by supporting a wider variety of statistical reports to a broader user community. Baseline data show that reports are currently provided to users within 11.5 days of request submission. Performance data for Web-enabled reporting are needed.

6) The STARS auto-entry project eliminated manual data entry for reports created by the MSHP. Performance measures indicate that the MSHP reports are available in STARS within 7 days versus a baseline of 22.5 days.

7) The MSHP GPS auto-population project supports automatic capture of latitude/longitude coordinate data via in-vehicle computers and GPS devices.

Performance measure is the percentage of crashes that "land" automatically on the GPS map.

8) The purpose of the STARS Accident Report Revision project is to increase MMUCC compliance of the official crash report form by increasing the number of fully compliant data elements and adding attributes to the partially compliant data elements.

9) The Document Management Workflow System at the Department of Revenue serves to reduce the time it takes to identify and sanction problem drivers by reducing the lag times in processing conviction, cord orders, title and registration actions from days to minutes. Performance measure data were not supplied.

10) The Judicial Case Management System – Justice Information System has been implemented in all 115 State courts. This program establishes electronic transmission of conviction data from all courts and reduces the lag time between conviction and posting to the driver record.

11) The CASE.net project provides public access to State court records. Secure access to confidential information is provided to law enforcement, prosecutors, and the courts.

12) Support for Municipal Court Automation is being added to the State court JIS automated reporting. Uploads of traffic case information from the municipal courts is expected to improve timeliness of posting convictions and to support a statewide tracking system. The timeline for adding municipal courts to the statewide court automation project is longer than for the initial JIS implementation.

13) The Missouri Ambulance Reporting System (MARS) is a NEMSIS-compliant EMS run report system composed of a centralized database and field data collection system for use by EMS providers. The State has measured timeliness and compliance with the NEMSIS standard, but the system should improve accuracy, integration, and accessibility as well. Accessibility is given a "green" indicator because the database of records does exist and is accessible to the EMS providers and the State health agency. The main limitation of MARS is that EMS providers are not required to report all traffic-related cases to the system. There are currently problems integrating this data with other sources of injury surveillance and crash data, but the improvements in timeliness and accessibility are documented.

NEBRASKA

Nebraska submitted five project descriptions that are compiled in the below.

System Quality Attribute	Crash	Roadway	Driver	Vehicle	Citation/ Adjudication	Injury Surveillance
Timeliness	X^1		$X^{2,3}$		X^4	
Accuracy						
Completeness			X^2			
Consistency						
Integration						
Accessibility						

Notes:

1) The project to Define and Implement Acceptance of Electronic Crash Data is designed to support electronic transfer of crash report data from law enforcement agencies to the Nebraska Department of Revenue. The project is in ongoing implementation, with additional agencies added to the process, as they are able. From a baseline of 120 days delay from crash event to data accessibility on the NDOR system, the project has reduced overall delay to below 90 days (preliminary 2009 data).

2) The Douglas County Court and the Nebraska DMV established a joint process to track fail-to-pay cases and to improve collection at the local level prior to referral to DMV for suspension. In the 6-month baseline period, Douglas County referred 4,141 cases to the DMV for failure to pay fines. In the 6 months following implementation, referrals totaled only 213 cases, a 93 percent reduction.

3) Acceptance of electronic death certificate data to post to the driver record reduced processing from a minimum of 90 days (paper reports were sent quarterly for data entry) to overnight, with 100 percent of records updated by the DMV within 2 days.

4) The Nebraska Crime Commission is managing a project to expand the citation automation program and the Statewide Citation File. Follow-up is needed to obtain a more complete description of this project and the ongoing e-citation efforts. The performance measure shows prosecutors receiving citation data within 2 days of issuance compared to a baseline of 14 days; but, it is not clear in the project scope whether data reported are for a local jurisdiction or statewide.

5) The EMS Data Quality Assessment and Improvement project supports electronic submission of EMS run reports to the central repository at the Department of Health and Human Services. DHSS encourages use of electronic field data collection and transmission by providing software to EMS agencies. 54 percent of providers submit data electronically, leading to a reduction in reporting delays from a baseline of 77 days to 11 days in 2008.

Appendix B: State Review by Regions

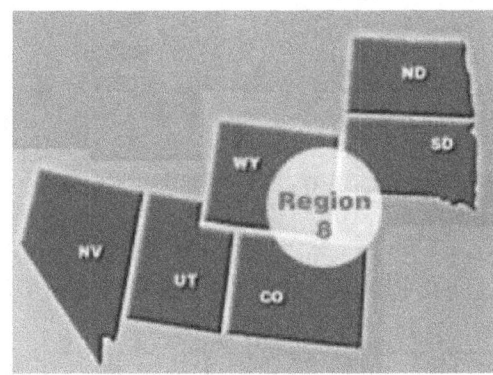

Region 8: Colorado, Nevada, North Dakota, South Dakota, Utah, Wyoming

Responses were received from Nevada, North Dakota, Utah, and Wyoming

NEVADA

Nevada provided a status report for its efforts to implement electronic crash and citation field data collection and to upgrade its central repository – the Nevada Citation and Accident Tracking System (NCATS). The following table was generated based on the progress report

System Quality Attribute	Crash	Roadway	Driver	Vehicle	Citation/ Adjudication	Injury Surveillance
Timeliness	X^1				X^1	
Accuracy	X^1				X^1	
Completeness						
Consistency						
Integration						
Accessibility	X^2				X^2	

Notes:

1) Since 2005, Nevada has doubled the number of agencies collecting crash and citation data electronically. In 2009 17 out of 36 agencies (covering >96 percent of the population) collect crash data and issue citations electronically.

2) The State's central repository – the Nevada Citation and Accident Tracking System (NCATS) – is collecting data electronically from the 17 participating agencies, covering approximate 88 percent of all crash reports

NORTH DAKOTA

North Dakota provided two project descriptions. The following table is derived from those descriptions.

System Quality Attribute	Crash	Roadway	Driver	Vehicle	Citation/ Adjudication	Injury Surveillance
Timeliness	X[1]				X[1]	
Accuracy	X[1,2]				X[1]	
Completeness	X[1]				X[1]	
Consistency	X[1]				X[1]	
Integration	X[2]	X[2]				
Accessibility						

Notes:

1) The State's implementation of TraCS began in 2006. At present, 57 percent (63 of 111) of law enforcement agencies are using TraCS and their reports account for 80 percent of all crash reports submitted to the State. The citation module was deployed more recently and is being used by 12 agencies. In 2010, the TraCS rollout to new agencies was put on hold as the State upgrades to TraCS v.10. Performance measures were not reported to support the project's impact on crash and citation data quality.

2) The Crash Location Conversion Project converted seven years of historical data in the Crash Reporting System from node-based location coding to a coordinate-based coding for use in GIS. The results were increased accuracy of location information for crashes and improved ability to link crash and roadway data to support safety analysis. The crash location conversion project is now complete with 100 percent coding of 8 years of data. No performance measures are used except the amount of data converted.

Appendix B: State Review by Regions

UTAH

Utah submitted the following table and supporting project descriptions:

System Quality Attribute	Crash	Roadway	Driver	Vehicle	Citation/ Adjudication	Injury Surveillance
Timeliness	X^1					X^3
Accuracy						
Completeness						X^3
Consistency	X^1					X^3
Integration					X^2	
Accessibility	X^1					X^3

Notes:

1) Utah rebuilt its electronic crash reporting system in 2006, deploying a Web-based system – the Centralized Crash Repository. From 2008 on, Utah has maintained data backlogs of 60 days or less from crash event to the date the data are available on the system. At the same time, a form revision resulted in an increase from 54 to 81 data elements in full compliance with the MMUCC guideline. The new Centralized Crash Repository also improves accessibility by supporting electronic uploads and downloads of crash data for authorized agencies, including several local law enforcement agencies.

2) The Administrative Office of the Courts has established a centralized database for court records. The plan is to convert all courts to the new system by July 1, 2011.

3) The Emergency Medical Services Bureau has updated its prehospital data reporting system (POLARIS) to support real-time incident reporting, meet current NEMSIS standards, improve data quality, and support better access for agencies submitting data to the statewide system. Performance measures for accessibility and timeliness were provided in the description. Some 75 percent of EMS providers are reporting to POLARIS electronically. There are now 139 agencies accessing data in Polaris. Timeliness of data is now 33 days post event, down from the baseline of 180 days in 2006, but an increase from the 21 days reported in 2007.

Appendix B: State Review by Regions

WYOMING

Wyoming provided seven project descriptions. The following table is derived from those descriptions.

Quality Attribute \ System	Crash	Roadway	Driver	Vehicle	Citation/ Adjudication	Injury Surveillance
Timeliness	X^2					
Accuracy		X^7				
Completeness	X^2					
Consistency	X^1					
Integration	$X^{3,5}$	$X^{5,6}$				
Accessibility	$X^{3,4,5}$	$X^{5,6,7}$				

Notes:

1) In January 2008, Wyoming implemented a new crash reporting form that is 98 percent MMUCC-compliant.

2) Wyoming implemented an electronic field data collection system for crashes. The system has improved timeliness from 83 days down to 16.2 days from date of the crash to availability of data on the system. In addition, the software includes error checks to ensure that key fields are not left blank on the form, improving completeness; but this latter is not measured with a data quality metric.

3) The Wyoming DOT has migrated historic crash data going back to 1994 to the new crash reporting database and format. This has allowed analysts to integrate data from all years into a single study without performing data compatibility clean up each time. The steps improved integration and accessibility. Performance measures are quantitative, but simply measure the number of years of data that are accessible.

4) Creation of an Oracle version of the existing SQL-based crash database has improved accessibility for users who prefer Oracle and are familiar with the Oracle reporting tools. No performance measures were supplied.

5) The CARE analysis system was deployed in 2008. It provides users with easy access to integrated roadway and crash data

6) Up to 20 roadway-related datasets are now available via CARE. The quantification of the number of datasets is the only performance measure.

7) Deployment of TRADAS has improved access to and accuracy of traffic count data used in safety analyses. Data quality metrics were not supplied.

Appendix B: State Review by Regions

The State included the following statements, but they have not been indicated in the table above due to a lack of detailed information:

- Driver/Integration – a snapshot was taken of the driver data that is based on a mainframe, and a set of Oracle tables was produced. This allows direct linkage between driver data and crash data, which is also in a set of Oracle tables

- Citation/Adjudication/Timeliness/Integration/Accessibility – The main improvement here was based on expanded deployment of *Full Court*, which provides a citation warehouse functionality for those courts that use it. While not all courts in Wyoming are yet aligned with Full Court, significant progress has been made pending the deployment of an electronic citation system: it is used in 15 of 23 county courts, all 29 circuit courts, and 8 municipal courts.

- Injury Surveillance/Accessibility – The main improvement here is through increased availability of information on the Web. A project that is pending final vendor selection/contract completion will usher in a major improvement in terms of trauma records (with significant improvements in timeliness, accuracy, completeness, and consistency).

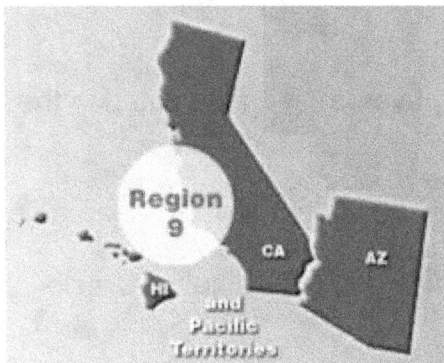

Region 9: Arizona, California, Hawaii, and Pacific Territories

No responses were received from these States.

Appendix B: State Review by Regions

Region 10: Alaska, Idaho, Montana, Oregon, Washington

Responses were received from Alaska and Oregon

ALASKA

Alaska submitted the following table and supporting project descriptions:

Quality Attribute \ System	Crash	Roadway	Driver	Vehicle	Citation/ Adjudication	Injury Surveillance
Timeliness	X[1,4,5]	X[5]	X[4,5]	X[5]	X[5]	X[1,4,5]
Accuracy	X[1,5]	X[5]	X[5]	X[5]	X[5]	X[1,5]
Completeness	X[1,5]	X[5]	X[5]	X[5]	X[5]	X[1,5]
Consistency	X[1,5]	X[5]	X[5]	X[5]	X[5]	X[1,5]
Integration	X[1,2,3]	X[2,3]	X[2,3]	X[2,3]	X[2,3]	X[1,2,3]
Accessibility	X[1,2,3]	X[2,3]	X[2,3]	X[2,3]	X[2,3]	X[1,2,3]

Notes:
From the submission, it was not clear how the described projects deliver all of the data quality improvements noted in the State's table. Performance measures were not supplied.

1) The Alaska Crash Outcomes Pilot Project is designed to merge crash, trauma registry, and hospital discharge data and to make the data available for analyses.

2) The Collaborative Statewide Governance for Criminal Justice and Highway Safety Data Sharing is to meet the need for an executive oversight committee of the TRCC and to incorporate the TraCS Steering Committee and the Multi-Agency Justice Integration Consortium (MAJIC) into the overall framework. Letters of agreement were signed by the 10 participating agencies.

3) The Traffic Records System Single Portal Pilot Project is to incorporate all geo-locatable data under one GIS to support spatial analysis and data integration.

4) The project to document digital cell phone signal strength on highways is designed to develop maps of coverage gaps for automatic crash notification and EMS communication systems, and to overlay those maps with crash locations.

5) The project to document latitude and longitude on crashes was designed to provide law enforcement with GPS units and to modify the State crash database to collect coordinates.

Appendix B: State Review by Regions

OREGON
Oregon submitted the following table and supporting project summaries:

System Quality Attribute	Crash	Roadway	Driver	Vehicle	Citation/ Adjudication	Injury Surveillance
Timeliness						
Accuracy						
Completeness	X[1]	X[5]				X[7]
Consistency						X[8]
Integration		X[4]			X[6]	X[9]
Accessibility	X[2,3]					

Notes:

1) The Map-Based Location Coding project enabled non-State highway crashes to be geo-locatable. Prior to 2007, only State highway crashes in the Statewide Crash Database contained an LRS value, which allows crash records to be geo-coded. Because of this project, more complete crash data is accessible via the TransGIS online query tool. A geo-coded crash has 68 data fields that display when queried. The addition of latitude and longitude coordinates to all crash data (State-maintained and local roads) was completed as of 2007. The 2007 value of 100 percent geocoded locations was an improvement over the 38.3 percent achieved in 2006.

2) Pd' [sic] Programming's Crash Magic software was purchased in June 2009. This software is used to create collision diagrams within ODOT and will have the capability to be used by cities and counties. Currently, the software is being tested and users are being trained to use this new tool. In 2010, comprehensive testing and configuration of the Crash Magic Tool continued. Oregon's file has crash elements not included in Crash Magic's standard configuration. With the primary objective of deploying a collision diagramming Web application that will benefit all jurisdictions, both State and local, additional time and effort is going into identifying and developing those required "high priority" functions. The application training goal is to train "power users" this summer. Each designated "power user" will in turn provide training assistance and support for the more infrequent users. Selected local government stakeholder will be part of the initial training group

3) Oregon was able to increase the percentage of law enforcement agencies using online crash data system for data retrieval and statistical reports from 6.8 percent of agencies (12 out of 177 agencies) in 2007. There are currently 21 agencies (11.9 percent of agencies) using an electronic citation reporting and 15 agencies (8.5 percent of agencies) using an electronic crash reporting system. During FY2009, Oregon State Police began the implementation of electronic citation and crash forms statewide with the purchase of 50 laptops and APS and ReportBeam

software installation for use in their cars. Currently, Oregon State Police is in the process of installing an additional 85 laptops in FY2010. The APS software includes a tool for collision diagramming, GPS, and electronic transfer of citation data from the system to local courts. ReportBeam software allows the user to query or map high incident areas. The rollout of ReportBeam software for field data collection of crash and citation data continues as follows: During 2009-2010, the primary focus for rolling out electronic crash and citation reporting has been at Oregon State Police (OSP). The funding for this OSP project was 100 percent non-408. OSP plans to implement this project to all traffic teams including in-car equipment by the end of 2010. The State anticipates additional law enforcement agency involvement in 2011.

4) Currently ODOT provides a screening of all State highways, producing a priority safety report of roadway segments using Safety Priority Index Systems (SPIS). To meet the Federal mandate of providing top 5 percent safety sites on all public roads, ODOT is undertaking expanding the SPIS to all public roads via a Geographic Information System platform. This project will provide reports and maps for safety screening for all public roads with traffic volume data (counts). ODOT is working to get traffic volume data to the same platform as location and crash data. If the data on traffic volumes are not available, we cannot duplicate the SPIS process. In June 2010, the State provided the following update: "The Pilot GIS/SPIS project continues: The completion of this project is dependent on getting traffic volumes for all public roads on the same platform, of which Oregon is now implementing a solution. The solution required some data manipulation, as well as adding a foreign key to the traffic volume database for all functionally classified roads (other than State highways). We are now in a position to join all the data on the same platform and we piloted the system to make sure it would work. We have hired an outside expert in GIS to give us recommendations on the architecture of the GIS platform. Once that recommendation is complete (June 2010), we will begin programming the system. This project is scheduled to be completed in April 2011."

5) In Oregon, many automatic traffic recorders (ATRs) were not capable of collecting speed data or length of vehicles. A few did not have multiple loops in each lane, so they did not have the infrastructure to collect the data. This project funded the purchase of Diamond traffic counters. RFP process was accomplished in the spring of 2009. Installing counters and software was primarily in the summer of 2009. Over 100 sites were outfitted to collect speed and length data.

6) Through the eCitation Project, Oregon increased the number of traffic citations distributed electronically from law enforcement agencies to local courts from approximately 33,000 citations in 2007 to 55,447 in FY 2009. A pilot was started in 2006 with Clackamas County Sherriff's Office to purchase handhelds, electronic citation and crash forms, and mobile printers. Currently, there are 21 agencies in Oregon (11.9 percent of agencies) using an electronic citation reporting system, with two more agencies planned during FY2010. In June 2010, the State provided the following update: "The project to collect speed data at

permanent ATR count stations continues: Oregon now has many more sites equipped to report speed and length data. We still have ten sites that cannot, because they do not have more than one loop in each lane, and there are seven sites located at signals that only collect volume data. Therefore, the total not collecting speed and length data is 17 sites. All other sites in Oregon have been outfitted with new Diamond Phoenix counters. Oregon now has about 90 percent of our permanent ATR sites that can report speed. Two years ago, there were 90 ATRs that were not reporting speed data and currently there are 17 sites. Oregon has been able to collect more complete traffic volume, speed, and classification data by reducing the number of ATRs that do not collect speed data."

7) The completeness of the statewide EMS/Injury Surveillance database has improved, as evidenced by the increase, from zero records in May 2008 to 24,089 records in January 2009, in the number of EMS run reports populating a pilot test database of run reports from May 2008 supplied by 19 service provider agencies. In June 2010, the State provided the following update: "EMS run reporting into the central database continues to increase. The EMS and Trauma Systems Program has contracted with the Oregon Fire Marshal's Office to continue to make the EMS database available to Oregon EMS agencies. As of June 2010, 39 ambulance services are entering data into the EMS patient encounter database, which is an increase in the number of NEMSIS-compliant agencies using the new system. When the contract becomes operational, the program will continue in cooperation with the Fire Marshal's Office for at least the five-year duration of its contract. When this becomes operational, an effort will be made to gather information from more EMS agencies."

8) The uniformity of the statewide EMS/Injury Surveillance database has improved, as evidenced by the increase, from zero records in May 2008 to 24,089 records in January 2009, each with 60 NEMSIS-compliant elements, in the number of EMS run reports populating a pilot test database of run reports from May, 2008 supplied by 19 service provider agencies. By June of 2010, 39 agencies were supplying run report data.

9) This is part two of a project that was started in 2008 using ImageTrend software to demonstrate the feasibility, challenges, and benefits of linking EMS patient records to hospital outcomes and other existing unique databases in the State (e.g., Oregon State Trauma System Database, Oregon State Hospital Discharge Database, ODOT Statewide Crash Database, Medical Examiner data). Data analysts were able to identify and describe which prehospital variables were most critical for matching and linking records with the EMS patient encounter database. Project deliverables included a summary document detailing a vision and general mechanics for development of a State EMS database linked to hospital outcomes.